TRANSPORTATION ISSUES, POLICIES AND R&D

UNMANNED AERIAL SYSTEMS

EARLY INTEGRATION PROGRESS AND CONSIDERATIONS

TRANSPORTATION ISSUES, POLICIES AND R&D

Additional books in this series can be found on Nova's website under the Series tab.

Additional e-books in this series can be found on Nova's website under the e-book tab.

TRANSPORTATION ISSUES, POLICIES AND R&D

UNMANNED AERIAL SYSTEMS

EARLY INTEGRATION PROGRESS AND CONSIDERATIONS

WILFRED SPARKS
EDITOR

nova publishers
New York

NOTICE TO THE READER

Library of Congress Cataloging-in-Publication Data

ISBN: 978-1-63484-652-3

Published by Nova Science Publishers, Inc. † New York

Contents

Preface		**vii**
Chapter 1	Unmanned Aerial Systems: FAA Continues Progress toward Integration into the National Airspace *United States Government Accountability Office*	**1**
Chapter 2	Statement of Margaret Gilligan, Associate Administrator for Aviation Safety, Federal Aviation Administration. Hearing on "Unmanned Aircraft Systems: Key Considerations Regarding Safety, Innovation, Economic Impact, and Privacy"	**49**
Chapter 3	Testimony of John Villasenor, Nonresident Senior Fellow, the Brookings Institution. Hearing on "Unmanned Aircraft Systems: Key Considerations Regarding Safety, Innovation, Economic Impact, and Privacy"	**57**
Chapter 4	Testimony of Paul Misener, Vice President of Global Public Policy, Amazon, Inc. Hearing on "Unmanned Aircraft Systems: Key Considerations Regarding Safety, Innovation, Economic Impact, and Privacy"	**75**

Chapter 5	Statement of Jeff VanderWerff, VWO, LLC., Representing the American Farm Bureau Federation. Hearing on "Unmanned Aircraft Systems: Key Considerations Regarding Safety, Innovation, Economic Impact, and Privacy"	**83**
Chapter 6	Summary of Subject Matter for the Hearing on "Ensuring Aviation Safety in the Era of Unmanned Aircraft Systems" *Subcommitte on Aviation Staff*	**87**
Chapter 7	Statement of Michael G. Whitaker, Deputy Administrator, Federal Aviation Administration. Hearing on "Ensuring Aviation Safety in the Era of Unmanned Aircraft Systems"	**93**
Chapter 8	Statement of James Hubbard, Deputy Chief, United States Forest Service. Hearing on "Ensuring Aviation Safety in the Era of Unmanned Aircraft Systems"	**101**
Chapter 9	Statement of Captain Tim Canoll, President, Air Line Pilots Association. Hearing on "Ensuring Aviation Safety in the Era of Unmanned Aircraft Systems"	**109**
Chapter 10	Statement of Richard Hanson, Director of Government and Regulatory Affairs, Academy of Model Aeronautics. Hearing on "Ensuring Aviation Safety in the Era of Unmanned Aircraft Systems"	**127**
Chapter 11	Testimony of Mykel Kochenderfer, Assistant Professor of Aeronautics and Astronautics, Stanford University. Hearing on "Ensuring Aviation Safety in the Era of Unmanned Aircraft Systems"	**157**
Chapter 12	Presidential Memorandum: Promoting Economic Competitiveness While Safeguarding Privacy, Civil Rights, and Civil Liberties in Domestic Use of Unmanned Aircraft Systems *Barack Obama*	**163**
Index		**169**

PREFACE

Unmanned aerial systems (UAS) are aircraft that do not carry a pilot aboard, but instead operate on pre-programmed routes or are manually controlled by following commands from pilot-operated ground control stations. Unauthorized UAS operations have, in some instances, compromised safety. The Federal Aviation Administration (FAA) Modernization and Reform Act of 2012 directed FAA to take actions to safely integrate UASs into the national airspace. In response, FAA developed a phased approach to facilitate integration and established test sites among other things. This book addresses the status of FAA's progress toward safe integration of UASs into the national airspace; research and development support from FAA's test sites and other resources; and how other countries have progressed toward UAS integration into their airspace for commercial purposes.

In: Unmanned Aerial Systems
Editor: Wilfred Sparks

ISBN: 978-1-63484-652-3

Chapter 1

UNMANNED AERIAL SYSTEMS: FAA CONTINUES PROGRESS TOWARD INTEGRATION INTO THE NATIONAL AIRSPACE*

United States Government Accountability Office

WHY GAO DID THIS STUDY

UASs are aircraft that do not carry a pilot aboard, but instead operate on pre-programmed routes or are manually controlled by following commands from pilot-operated ground control stations. Unauthorized UAS operations have, in some instances, compromised safety. The FAA Modernization and Reform Act of 2012 directed FAA to take actions to safely integrate UASs into the national airspace. In response, FAA developed a phased approach to facilitate integration and established test sites among other things.

GAO was asked to review FAA's progress in integrating UASs. This report addresses (1) the status of FAA's progress toward safe integration of UASs into the national airspace, (2) research and development support from FAA's test sites and other resources, and (3) how other countries have progressed toward UAS integration into their airspace for commercial purposes. GAO reviewed and analyzed FAA's integration- planning

* This is an edited, reformatted and augmented version of the United States Government Accountability Office publication, GAO-15-610, dated July 2015.

documents; interviewed officials from FAA and UAS industry stakeholders; and met with the civil aviation authorities from Australia, Canada, France, and the United Kingdom. These countries were selected based on several factors including whether they have regulatory requirements for commercial UASs, operations beyond the view of the pilot, and whether non-military UAS are allowed to operate in the airspace. In comments on this report, the Department of Transportation noted that GAO did not address environmental considerations of UAS integration; such considerations were outside the scope of this report.

WHAT GAO FOUND

The Federal Aviation Administration (FAA) has progressed toward its goal of seamlessly integrating unmanned aerial system (UAS) flights into the national airspace. FAA has issued its *UAS Comprehensive Plan* and *UAS Integration Roadmap*, which provide broad plans for integration. However, according to FAA, it is working with MITRE to develop a foundation for an implementation plan; FAA then expects to enact a plan by December 2015. While FAA still approves all UAS operations on a case-by-case basis, in recent years it has increased approvals for UAS operations. Specifically, the total number of approvals for UAS operations has increased each year since 2010, and over the past year has included approvals for commercial UAS operations for the first time. In addition, FAA has issued a Notice of Proposed Rulemaking that proposes regulations for small UASs (less than 55 pounds).

The FAA's six designated test sites have become operational but have had to address various challenges during the process. The designated test sites became operational in 2014, and as of March 2015, over 195 test flights had taken place. These flights provide operations and safety data to FAA in support of UAS integration. In addition, FAA has provided all test sites with a Certificate of Waiver or Authorization allowing small UAS operations below 200 feet anywhere in the United States. However, during the first year of operations, the test sites faced some challenges. Specifically, the test sites sought additional guidance regarding the type of research they should conduct. According to FAA, it cannot direct the test sites, which receive no federal funding, to conduct specific research. However, FAA did provide a list of potential research areas to the test sites to provide some guidance on areas for research. FAA has conducted other UAS research through agreements with MITRE and some universities, and in May 2015 named the location of the

UAS Center of Excellence—a partnership among academia, industry, and government conducting additional UAS research. Unlike FAA's agreements with the test sites, many of these arrangements have language specifically addressing the sharing of research and data.

Around the world, countries have been allowing UAS operations in their airspace for purposes such as agricultural applications and aerial surveying. Unlike in the United States, countries GAO examined—Australia, Canada, France, and the United Kingdom—have well-established UAS regulations. Also, Canada and France currently allow more commercial operations than the United States. While the United States has not finalized UAS regulations, the provisions of FAA's proposed rules are similar to those in the countries GAO examined. However, FAA may not issue a final rule for UASs until late 2016 or early 2017, and rules in some of these countries continue to evolve. Meanwhile, unlike under FAA's proposed rule, Canada has created exemptions for commercial use of small UASs in two categories that allow operations without a government-issued certification, and France and Australia are approving limited beyond line-of-sight operations. Similar to the United States, other countries are facing technology shortfalls, such as the ability to detect and avoid other aircraft and obstacles, as well as unresolved issues involving limited spectrum that limit the progress toward full integration of UASs into the airspace in these countries.

ABBREVIATIONS

ASTM	International formerly known as the American Society for Testing and Materials
COA	Certificates of Waiver or Authorization
DHS	Department of Homeland Security
DOD	Department of Defense
DOT	Department of Transportation
EASA	European Aviation Safety Agency
EUROCAE	European Organization for Civil Aviation Equipment
Experimental Certificate	Special Airworthiness Certificates in the Experimental Category
FAA	Federal Aviation Administration
ICAO	International Civil Aviation Organization
MITRE	MITRE Corporation

MOA	memorandum of agreement
MOU	memorandum of understanding
NASA	National Aeronautics and Space Administration
NPRM	Notice of Proposed Rule Making
RTCA	formerly the Radio Technical Commission for Aeronautics (now RTCA)
the 2012 Act	FAA Modernization and Reform Act of 2012
UAS	unmanned aerial systems or unmanned aircraft systems

* * *

July 16, 2015

Congressional Committees

Since the early 1990s, unmanned aerial systems (UAS) have operated on a limited basis in the national airspace system, primarily supporting public uses, such as military and border-security operations.[1] The list of potential uses is now rapidly expanding to include a broad range of UAS operations, including assisting in search and rescue operations, inspecting pipelines, photographing real estate, surveying land and crops, disaster assistance, gathering news, and filming movies. According to a 2013 report by a UAS industry group, the economic benefit of fully integrating UASs into the national airspace system will total more than $13.6 billion in the first 3 years of integration and grow to more than $82.1 billion from 2015 through 2025.

The Federal Aviation Administration's (FAA) goal is to seamlessly integrate all UAS operations into the national airspace. While FAA is developing regulations to allow routine UAS commercial operations, it currently only allows them in the national airspace after a case-by-case safety review and approval. At the same time, the safety of the national airspace is threatened on a nearly daily basis by UASs' operating without approval. For example, one UAS nearly collided with a New York Police Department helicopter over New York City, another came dangerously close to a US Airways regional jet over the Florida panhandle, and numerous UASs have been spotted flying over professional and college football stadiums full of people. The FAA has stated the number of reported incidents has increased recently, with 97 incidents reported from February through March 2015, compared to just 3 incidents reported during the same period in 2014. FAA believes the increase is due to more awareness about UASs and the need to report incidents, but the extent these factors contribute to the increase is not clear.

As UAS operations in the national airspace continue to increase, members of Congress have raised questions about the safe and efficient integration of UASs into the national airspace system and asked us to examine the progress of UAS integration. This report examines (1) the status of FAA's progress toward safe integration of UASs into the national airspace, (2) research and development support from FAA's test sites and other resources, and (3) how other countries have progressed toward UAS integration into their airspace for commercial purposes.

To address all the objectives, we reviewed FAA reports and prior GAO work[2] related to integrating UASs in the national airspace system, reviewing such work as FAA's framework for UAS integration and our prior work on UAS research and development, efforts by FAA to accommodate ongoing research and commercial UAS use, and UASs' technology challenges.[3] We also reviewed provisions of the FAA Modernization and Reform Act of 2012[4] (the 2012 Act), and the Notice of Proposed Rulemaking for small UAS operations.[5] To determine FAA's progress toward safe integration of UASs into the national airspace, we reviewed and analyzed documents provided by various stakeholders and conducted semi-structured interviews with these stakeholders, including government—FAA, Department of Defense (DOD), National Aeronautics and Space Administration (NASA), and the Department of Homeland Security—academics researching UAS technology, industry trade associations, and others from the UAS industry including a manufacturer and a commercial operator. We selected the private-sector stakeholders, such as RTCA and ASTM International, based on their involvement within UAS standards groups.[6] To identify research and development support from FAA's test sites and other resources, we reviewed and analyzed documents provided by officials at each of FAA's six test sites and conducted semi-structured interviews with officials from each of the six, including the applications submitted by the selected test sites and quarterly reports provided to FAA. We also interviewed FAA officials managing the test site program and those managing other research and development efforts related to UASs. To identify how other countries have progressed toward UAS integration into their airspace for commercial purposes, we obtained the UAS regulations of authorities in France, the United Kingdom, Australia, and Canada and interviewed the civil aviation authorities in them. We selected these countries based on several factors including whether they have regulatory requirements for commercial UASs, operations beyond the view of the pilot, and whether the country allows non-military UAS to operate in the airspace. We also reviewed key documents and interviewed international aviation entities including the International Civil Aviation Organization (ICAO).[7] More details on our scope and methodology can be found in appendix I.

We conducted this performance audit from January 2014 to July 2015 in accordance with generally accepted government auditing standards. Those standards require that we plan and perform the audit to obtain sufficient, appropriate evidence to provide a reasonable basis for our findings and conclusions based on our audit objectives. We believe that the evidence obtained provides a reasonable basis for our findings and conclusions based on our audit objectives.

BACKGROUND

UASs can be categorized by both size and mission, as shown in figure 1.

- For the purposes of this report, in terms of size, we use the broad categories of "small" and "large" UASs. Small UASs typically weigh less than 55 pounds and can be used for a variety of commercial purposes including photography and package delivery. According to an industry association, small UASs are expected to comprise the majority of UASs that will operate in the national airspace system. Large UASs, depending on their size and purpose, generally fly at higher altitudes and are used for the purposes of surveillance, data gathering, and communications relay.
- UAS operations are also categorized by how they are being used—their mission—within line of sight of the operator or beyond the line of sight of the operator. For UAS operations within the operator's line of sight—for example a real estate agent taking photographs of a house—the operator relies only on their vision to avoid colliding with other objects. On the other hand, UAS operations occurring beyond the line of sight of the operator—for example conducting rail or pipeline inspections—requires that FAA segregate the airspace or that

the UAS needs instruments to sense other aircraft and obstacles and avoid those obstacles, as well as, other technologies that will keep the aircraft safely operating during its mission.

The FAA plays two major roles in integrating UASs into the national airspace—a regulator and a service provider. As the regulator, the FAA seeks to ensure the safety of persons and property in the air and on the ground in part by requiring that UAS operators and manufactures follow specific operation and manufacturing standards. As the service provider, the FAA is responsible for providing safe and efficient air-traffic control services in the national airspace system and the other portions of global airspace. In addition to FAA, many federal and private sector entities have roles in the effort to integrate UAS into the national airspace system. See table 1 for UAS stakeholders and their responsibilities.

Table 1. Unmanned Aerial Systems (UAS) Stakeholders and Their Responsibilities

Agency	Responsibility
Federal Aviation Administration (FAA)	FAA's UAS Integration Office is responsible for ensuring that Unmanned Aerial Systems (UAS) operate safely in the national airspace system and for providing safe and efficient air traffic control.
Department of Defense (DOD)	DOD provides FAA with UAS operational and safety data, as well as research and development support.
National Aeronautics and Space Administration (NASA)	NASA provides research and development and testing on UAS integration efforts.
Department of Homeland Security (DHS)	DHS's Customs and Border Protection has provided flight demonstrations to FAA's Next Generation Air Transportation System (NextGen) Office.
UAS Executive Committee[a]	The UAS Executive Committee is composed of senior executives from federal agencies including FAA, DOD, NASA, and DHS and is responsible for identifying solutions to the range of technical, procedural, and policy concerns arising from UAS integration.

Table 1. (Continued)

Agency	Responsibility
UAS Aviation Rulemaking Committee	The UAS Aviation Rulemaking Committee was chartered in 2011 to provide a mechanism for industry and academic stakeholders as well as other federal, state, and local government entities to provide recommendations and standards to FAA on issues related to UAS integration.[b]
RTCA Special Committee 228	RTCA is a private, not-for-profit organization consisting of industry experts. Special Committee 228 is responsible for developing standards for UASs to avoid objects and other aircraft, and supporting the radio controls of a UAS.
ASTM International Committee F38[c]	ASTM International Committee F38 is a committee consisting of industry experts that are responsible for developing standards and consensus based recommendations for small UAS integration into the national airspace system and worldwide.

Source: GAO. | GAO-15-610

[a] The UAS Executive Committee was formed as a result of the National Defense Authorization Act for Fiscal Year 2010 (Pub. L. No. 111-84, 123 Stat. 2190 (2009)). Section 935 of the 2010 National Defense Authorization Act states that "The Secretary of Defense and the Secretary of Transportation shall, after consultation with the Secretary of Homeland Security, jointly develop a plan for providing expanded access to the national airspace system for unmanned aircraft systems of the Department of Defense" and requires the Executive Committee members to provide Congress with, among other things, a communications plan, specific milestones for expanded access to the national airspace system, and a report on their efforts.

[b] FAA extended the Aviation Rulemaking Committee's charter through June 17, 2016, requesting the Committee continue to provide information, advice, and recommendations to the FAA for UAS integration.

[c] ASTM International, formerly known as the American Society for Testing and Materials, works to deliver the test methods, specifications, guides, and practices that support industries and governments worldwide.

FAA also partners with a range of industry, federal research entities, universities, and international organizations for research and development on UAS issues.

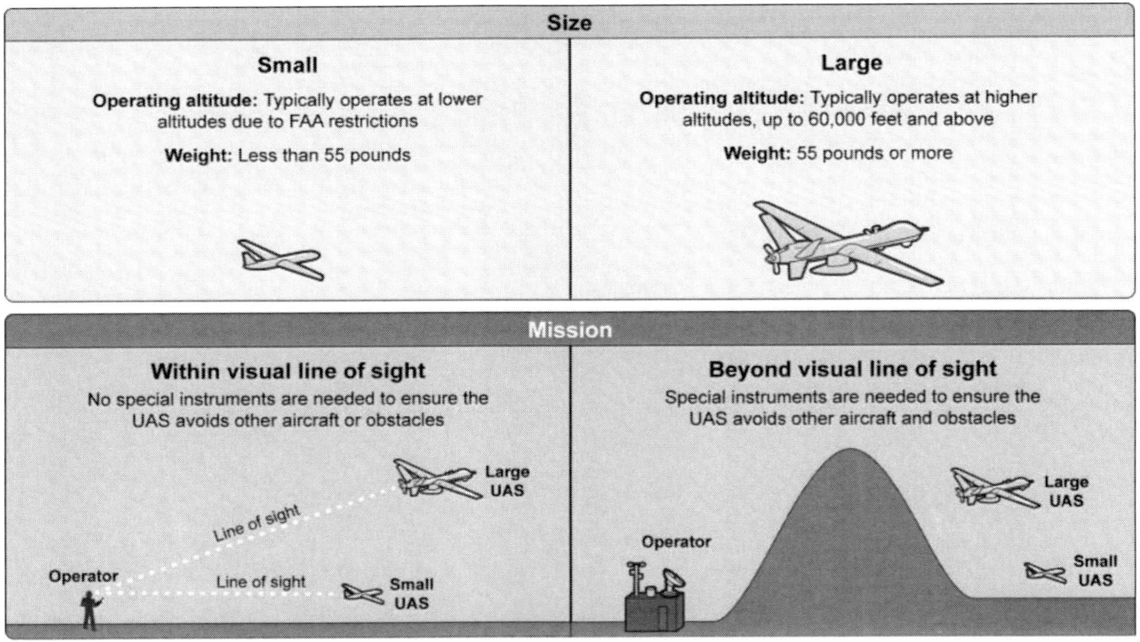

Source: | GAO-15-610

Figure 1. Unmanned Aerial Systems (UAS) Size and Mission Categories.

- Federally Funded Research and Development Centers[8] and Cooperative Research and Development Agreements typically require an agency, organization, or company to perform specific research and provide FAA with data in exchange for funding. FAA uses these types of agreements to support research and development in critical technologies needed for UAS integration including “detect and avoid” and “command and control” as well as for the dedicated radio-frequency spectrum and “human factors.”[9]
- Some other partnerships use Other Transaction Agreements to establish the research and development relationship.[10] However, these take many forms and are subject to requirements that may differ from the federal laws and regulations that apply to contracts, grants, or cooperative agreements, and therefore might not include a requirement to share research results with FAA.

Currently, most UAS operations must remain within visual line of sight of the UAS operator. FAA’s long-term goal is to pursue research and development that will advance technology in these critical areas, such as detect and avoid, and supporting beyond-visual-line-of-sight operations. These types of operations, according to an industry group, have the most potential for commercial purposes.

In response to the 2012 Act, FAA has been planning for UAS integration into the national airspace and has been taking steps toward increasing UAS operations.[11] The 2012 Act outlined 17 date specific requirements and set deadlines for FAA to achieve safe UAS integration by September 2015. These requirements included developing two planning documents-UAS Comprehensive Plan and the UAS Roadmap. FAA has completed these two requirements in addition to naming six test sites where research and development will occur.[12] However, we found in December 2014 that several other requirements, some key ones, including the publication of a final rule on small UAS operations, had not been completed (see app. II).[13]

As part of its role in supporting UAS integration, FAA authorizes all UAS operations (access to the airspace as well as the aircraft itself) in the national airspace system—military; public (academic institutions and, federal, state, and local governments including law enforcement organizations); and civil (non-government including commercial).[14] Depending on the type of user, public or civil, the process for accessing the airspace may be different (see table 2).

Table 2. Federal Aviation Administration (FAA) Approvals Required for Unmanned Aerial Systems to Access the National Airspace

Type of operations	Airspace approval	Equipment approval
Public (academic institutions and federal, state, and local governments including law enforcement organizations)	Public Certificates of Waiver or Authorization (COA)	Public COA
Civil (non-government, including commercial)	Civil COA	Section 333 exemption, Special Airworthiness Certificates in the Experimental Category

Source: GAO presentation of FAA information. | GAO-15-610

Currently, since a final rulemaking is not yet completed, FAA only approves UASs access to the national airspace on a case-by-case basis. FAA provides this approval through three different means:

- Public or Civil Certificates of Waiver or Authorization (COA): A COA is an authorization, generally for up to 2 years, issued by the FAA to an operator for a specific UAS activity. Public entities, including FAA- designated test sites (described in more detail later), and civil entities may apply for a COA to obtain authorized access to the airspace. FAA has a goal to review and approve all COAs within 60 days of being received.
- Section 333 exemptions: Since September 2014, commercial entities have applied to FAA for exemptions under section 333 of the 2012 Act, Special Rules for Certain Unmanned Aircraft Systems. This exemption requires the Secretary of Transportation to determine if certain UASs may operate safely in the national airspace system prior to the completion of UAS rulemakings.
- Special Airworthiness Certificates in the Experimental Category and the Restricted Category (Experimental Certificate): Civil entities, including commercial interests, may apply for experimental certificates, which may be used for research and development, commercial operations, training, or demonstrations by manufacturers.

While FAA has proceeded in planning for integration, foreign countries are also experiencing an increase in UAS use and planning for integration, and some have begun to allow commercial entities to fly UASs under limited circumstances. Some countries have already established regulations for flying UASs or formal processes for exemptions, while others have taken steps to completely ban all UAS operations. While some countries have worked independently to integrate UAS operations, some international groups, such as the ICAO, are working to harmonize UAS regulations and standards across borders.

FAA Has Made Progress in the First Phase of UAS Integration

Developing Planning Documents

FAA has taken a number of steps to move toward further safe integration of UAS in the national airspace in response to key requirements of the 2012 Act. FAA has developed the following planning documents:

- In November 2013, FAA issued the *UAS Comprehensive Plan*[15] identifying six high-level strategic goals for integrating UAS into the national airspace, including routine public operation and routine civil operations.[16] The *Comprehensive Plan* provides a phased-in approach for achieving these goals, which will initially focus on public UAS operations, but ultimately will provide a framework for civil UAS operations. According to the plan, each partner agency will work to achieve these national goals and may develop agency-specific plans that are aligned to the national goals and objectives. The DOD's *Unmanned Systems Integrated Roadmap* and FAA's *UAS Integration Roadmap*, described below, are examples of such plans.[17]
- In November 2013, FAA also issued the *UAS Integration Roadmap*, which identified a broad three phase approach to UAS integration—accommodation, integration, and evolution—with associated priorities for each phase.[18] These priorities provide insight into FAA's near-, mid-, and far-term goals for UAS integration, as shown in figure 2. FAA intends to use this approach to facilitate further incremental steps toward its goal of seamlessly integrating UAS flight into the

national airspace. Under this approach, FAA's initial focus during the accommodation phase will be on safely allowing for the expanded operation of UASs by selectively accommodating some use and demonstrating progress by increasing operations throughout the phase. In the integration phase, FAA plans to develop new operational rules, standards, and guidance, shifting its emphasis toward moving beyond case-by-case approval for UAS use, once technology can support safe operations. Finally, in the evolution phase, FAA plans to focus on revising its regulations, policy, and standards based on the changing needs of the airspace. This phased approach has been supported by stakeholders across both academia and industry. The 2012 Act requires the Roadmap to be updated annually, but as of May 2015 FAA has only issued one version of the Roadmap. FAA intends to update the *Roadmap* by September 2015 and send it to the Office of Management and Budget for additional review before it is publicly released.

While these planning documents provide a broad framework for integration, FAA is still in process of developing an implementation plan for integrating UASs. FAA's Comprehensive Plan and Roadmap provide broad plans for integration, but are not detailed implementation plans that predict with any certainty when full integration will occur and what resources will be needed. The Department of Transportation's Inspector General issued a report in June 2014 that recommended FAA develop an implementation plan.[19]

Two reports—one from the UAS Aviation Rulemaking Committee and a second internal FAA report—have discussed the importance of an implementation plan and information to include as part of such a plan. The UAS Aviation Rulemaking Committee has emphasized that FAA needs an implementation plan that would identify the means, necessary resources, and schedule to safely and expeditiously integrate civil UAS into the national airspace. The proposed implementation plan contains several hundred tasks and other activities needed to complete the UAS integration process, including:

- identifying gaps in current UAS technologies, regulations, standards, policies, or procedures;
- developing new technologies, regulations, standards, policies, and procedures that would support safe UAS operations; and

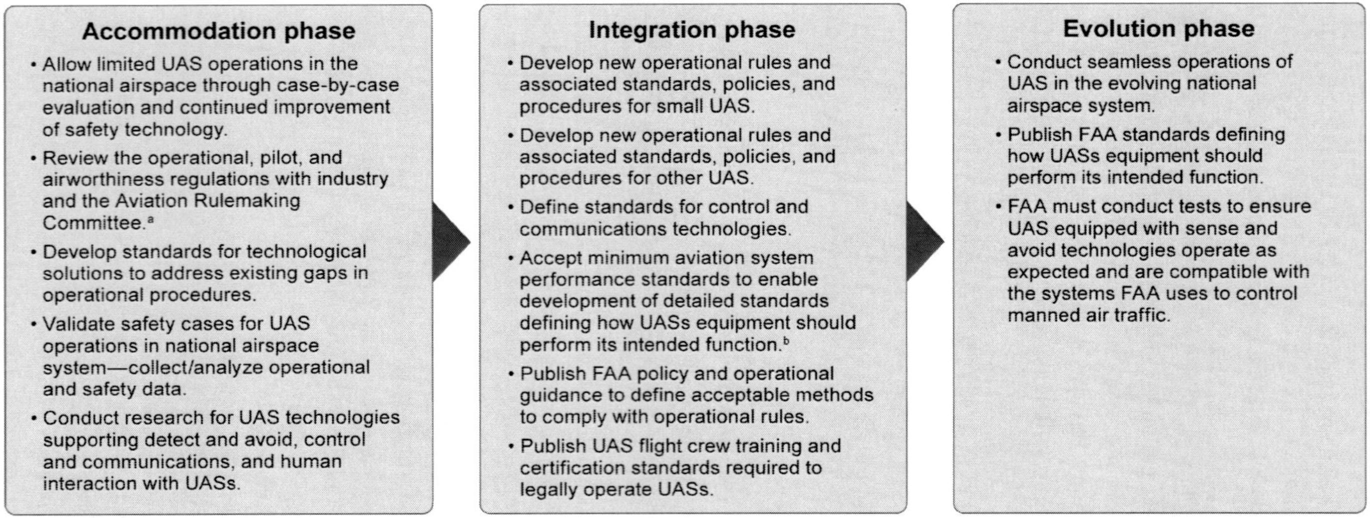

Source: Federal Aviation Administration (FAA) UAS Roadmap. | GAO-15-610

[a] FAA's aviation policies and regulations focus on overall safety being addressed through three primary areas: equipment (airworthiness), personnel (pilot), and operations. Each of these areas has standards and minimum levels of safety that must be met independent of each other. Since these policies and regulations were not designed with UAS in mind, they must be reviewed and revised to safely accommodate UAS in the national airspace.

[b] Minimum aviation-system performance standards specify characteristics that are useful to designers, installers, manufacturers, service providers, and users of systems intended for operational use within a defined airspace.

Figure 2. Priorities for Federal Aviation Administration's (FAA) Three Phases of Unmanned Aircraft Systems (UAS) Integration as Described in the *UAS Integration Roadmap*.

- identifying necessary activities to advance routine UAS operations in the national airspace, including the development of guidance materials, training, and procedures for certification of aircraft.

An internal FAA report from August 2014 prepared by MITRE Corporation (MITRE) was intended to assist FAA's development of the key components of an implementation plan.[20] This report suggested that among other actions FAA's implementation plan should:

- identify the tasks necessary, responsibilities, resources, and expected time frames for incremental expansion of UAS operations;
- clarify the priorities for aligning internal resources in support of near-term and long-term integration efforts and provide consistent communication with external stakeholders on the expected progress, cost, and extent of UAS integration during these time periods;
- align resources supporting UAS integration including allocation of FAA personnel and funds used for contracts and to acquire systems and services in support of integration; and
- establish the operational, performance, and safety data needed and also the associated infrastructure for collecting, storing, disseminating, and analyzing data, actions that could be a component of an implementation plan.

According to FAA, it continues to work with MITRE developing the foundation for a detailed implementation plan. FAA expects MITRE to complete this by September 2015 and FAA to enact the plan by December 2015.[21] According to FAA, the agency used the Aviation Rulemaking Committee's report when writing the *Roadmap* and is applying the report prepared by MITRE to help develop the detailed implementation plan.

Issuing a Final Rule for Small UAS Operations

In February 2015, FAA made progress toward the 2012 Act's requirement to issue a final rule for the operations of small UASs—those weighing less than 55 pounds—by issuing a Notice of Proposed Rulemaking (NPRM) that could, once finalized, allow greater access to the national airspace.[22] To mitigate risk, the proposed rule would limit small UASs to daylight-only operations, confined areas of operation, and visual-line-of-sight operations.

This proposed rule also addressed other issues pertinent to UAS operations including aircraft registration, operations in the national airspace, and operator certification. See table 3 for a summary of the rule's major provisions. FAA's release of this proposed rule for small UAS operations started the process of addressing remaining requirements of the 2012 Act.

Table 3. Selected Provisions of Proposed Rule for Small Unmanned Aircraft Systems (UAS) Operations, by Major Category

Category	Summary of proposed requirements
Operational limitations	• Must weigh less than 55 lbs. (25 kg). • Must operate within visual line of sight only. • May not operate above any persons not directly involved in the operation. • Must only operate during the day, no nighttime operations. • Maximum airspeed of 100 mph. • Maximum altitude of 500 feet above ground level. • Must not operate carelessly or recklessly. • Establishment of a micro-UAS category (4.4 lbs. or less). • Must yield right-of-way to other aircraft, manned or unmanned.
Operator certification and responsibilities	• Must pass a knowledge test initially and every 24 months. • Must be vetted by the Transportation Security Administration. • Must obtain an unmanned-aircraft operator's certificate with a small UAS rating.
Aircraft requirements	• Federal Aviation Administration (FAA) airworthiness certification not required, but operator must inspect the UAS to ensure that it is in a condition for safe operation. • Aircraft markings required; if aircraft is too small to display markings in standard size, then the aircraft simply needs to display markings in the largest practicable manner.
Model aircraft	• Would not apply to model aircraft that satisfy all of the criteria specified in section 336 of Public Law 112-95. • Would incorporate a 2012 Act provision that preserves the FAA's enforcement authority to pursue enforcement against model aircraft operators who endanger the safety of the national airspace system.

Source: Notice of Proposed Rulemaking for small UAS. | GAO-15-610

FAA's proposed rule also sought comments on a potential micro-UAS classification (4.4 pounds or less) that would apply to very small UAS being used for authorized purposes. This classification would be based on the UAS Aviation Rulemaking Committee's recommendations, as well as approaches adopted in other countries, that a separate set of regulations for micro-UASs be created. FAA is considering provisions for micro-UASs classification such as limiting operation airspeed to 30 knots, limiting flight to within visual line of sight, and having aircraft made out of materials that break on impact.

The proposed rule would not apply to model aircraft—unmanned aircraft that are flown for hobby or recreational purposes, capable of sustained flight in the atmosphere, and flown within visual line of site of the person who is operating the aircraft—as specified in the 2012 Act.[23] Whether or not a UAS is considered a model aircraft or a small UAS depends upon its operation. For example, if the operator is flying an unmanned aircraft for recreational purposes the unmanned aircraft is considered a model aircraft. If the exact same type of unmanned aircraft is being operated for an authorized purpose such as a search and rescue mission, it is considered a small UAS.[24] The 2012 Act specifically prohibits FAA from promulgating rules regarding model aircraft that meet specific criteria, including model aircraft flown strictly for hobby or recreational use and operated in a manner that does not interfere with and gives way to any manned aircraft.[25] However, the proposed rule would incorporate the 2012 Act provisions that preserve FAA's authority to pursue enforcement against persons operating model aircraft who endanger the safety of the national airspace system.[26]

According to FAA, it may take 16 months to process the comments it receives on the NPRM and develop and issue the final rule for small UAS operations. If FAA takes 16 months, the final rule would be issued in late 2016 or early 2017, about two years beyond the requirement in the 2012 Act. However, during the course of our work, FAA told us that the time needed to respond to a large number of comments could further extend the time to issue a final rule. When the comment period closed on April 24, 2015 FAA had received over 4,500 comments. FAA officials told us that it has taken a number of steps to develop a framework to efficiently process the comments it expects to receive. Specifically, they said that FAA has a team of employees assigned to lead the effort with contractor support to track and categorize the comments as soon as they are received.

Creating Test Sites

FAA has also met the requirement from the 2012 Act to create UAS test sites for research and development. Specifically, in December 2013 FAA selected six UAS test site locations, which all became operational between April 2014 and August 2014. According to FAA, these sites were chosen based on a number of factors including geography, climate, airspace use, and a proposed research portfolio that was part of the application. Under FAA policy, all UAS operations at a test site must be authorized by FAA through either the use of a COA or an experimental certificate. In addition, FAA does not provide funding to support the test sites. Thus, these sites rely upon revenue generated from entities, such as those in the UAS industry that are using the sites for UAS flights. The 2012 Act authorized the test sites to operate until February 14, 2017. FAA stated it is too early to assess the test sites' results and effectiveness and thus whether the test sites should be extended. According to FAA officials, FAA does not object to extending the test sites but may need additional resources if that happens.

Accommodating Limited UAS Operations

Although it still relies on case-by-case approvals, FAA has increased UAS operations during the accommodation phase of UAS integration. As we have previously noted, UAS operators can only gain access to the national airspace by obtaining a COA, an experimental certificate, or a section 333 exemption. From 2010 to 2014, the total number of COAs approved for public operations has increased each year, with FAA issuing 403 COAs thus far this year, as shown in table 4. Similarly, from 2011 to 2014, the total number of experimental certificates has increased each year, with FAA issuing 6 thus far this year.

Table 4. Total Number of Federal Aviation Administration-Issued Public Certificates of Waiver or Authorization (COA) and Special Airworthiness Certificates for Experimental Aircraft for Unmanned Aerial Systems

	2010	2011	2012	2013	2014	2015
Public COA	286	309	383	407	609	403[a]
Experimental certificate	15	14	18	26	39	6[b]

Source: FAA | GAO-15-610 [a] As of April 30, 2015. [b] As of May 1, 2015.

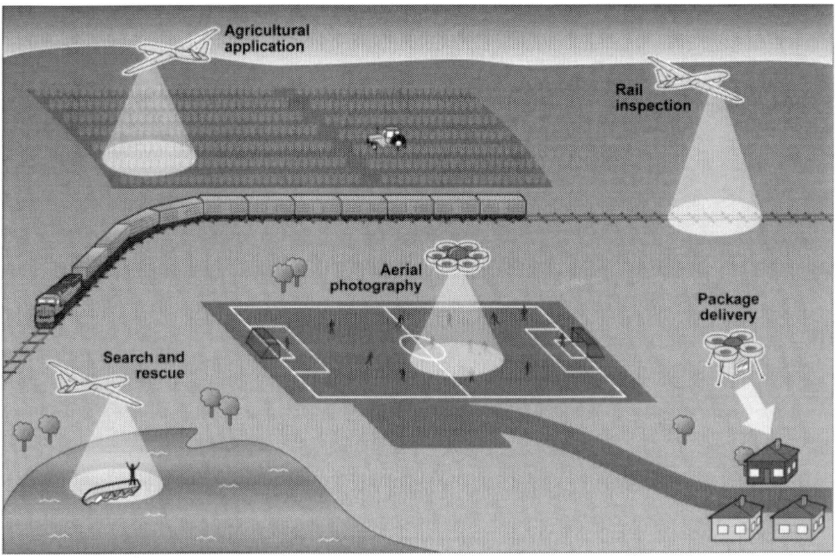

Source: GAO. | GAO-15-610

Figure 3. Examples of Potential Commercial Unmanned Aerial Systems Operations.

In September 2014, FAA granted the first section 333 exemptions; at that time 6 total exemptions were granted for commercial UAS operations to movie and TV production companies. As of June 9, 2015, FAA had granted 548 section 333 exemptions to companies for a variety of additional commercial operations supporting the real estate, utility, and agriculture industries, among others. See figure 3 for examples of commercial uses, including some approved under section 333 exemptions.

FAA has taken steps to make access easier for those operating UASs under a section 333 exemption. On March 23, 2015, FAA established an interim policy to speed up authorizations for certain commercial unmanned aircraft operators that request Section 333 exemptions. According to FAA, the new policy helps bridge the gap between the past "case-by-case" approval process, which evaluated every commercial UAS operation individually, and future operations after they publish a final version of the proposed small UAS rule. Under the new policy, the FAA will grant a COA for flights at or below 200 feet to any commercial UAS operator with a Section 333 exemption for aircraft that weigh less than 55 pounds, operate during the daytime, operate within visual line of sight of the pilots, and stay certain distances away from airports or heliports. According to FAA, the "blanket" 200-foot COA allows flights anywhere in the country except restricted airspace and other areas, such as major cities, where the FAA prohibits commercial UAS operations.[27] The agency expects the new policy will allow companies and individuals who want to use UAS within these limitations to start flying much more quickly than before. A company wanting to operate above 200 feet, or outside the other rules set up by FAA, must obtain a separate COA.

FAA took additional steps in May 2015 to work with industry to safely expand UAS operations. FAA announced its Pathfinder Program that will partner FAA with companies to perform research in support of UAS integration. These companies will focus on using UAS for specific applications, such as news gathering and surveying crops. In addition, two will focus on applications for beyond the visual line of sight of the operator. One industry stakeholder stated the next step would be to develop additional mechanisms to allow UAS operations beyond the visual line of sight of the operator once technology supports greater use.

POTENTIAL COMMERCIAL AGRICULTURE USE

The University of California-Davis has been conducting a demonstration project using a Yamaha RMAX helicopter to spray pesticides at its vineyard. The purpose of the project is to test the efficiency and safety of aerial spraying of grape crops. According to an official from Yamaha Motor Corporation, USA, initial tests have shown the RMAX can cover up to 12 times as much area in an hour compared with traditional spraying methods

Source: Yamaha Motor Corporation, USA | GAO-15-610

While accommodating UAS access, FAA and industry have taken steps to educate UAS operators on how to safely operate. UAS industry stakeholders and FAA have begun an educational campaign that provides prospective users with information and guidance on flying safely and responsibly. Specifically, they launched an informational website for UAS operators to ease public concerns about privacy and support safer UAS operations in the national airspace.[28] In addition, in May 2015, FAA announced plans to develop the "B4UFLY" smartphone application designed to help UAS users, both model aircraft and recreational UAS operators, know where it is safe and legal to fly. The application is designed to let an operator know if it is safe and legal to fly in a specific location.

Collaborating with UAS Stakeholders

FAA has worked with federal and industry stakeholders to coordinate federal activities in support of conducting research and development and creating UAS standards to facilitate UAS integration. As with other large government-wide initiatives, achieving results for the nation increasingly requires that federal agencies and others work together. FAA has worked with the UAS Executive Committee to facilitate federal UAS activities and RTCA Special Committee 228, ASTM International Committee F38, and the UAS Aviation Rulemaking Committee to develop safety, reliability, and performance standards for UAS.

Each collaborative group has defined different long-term goals in support of UAS integration and has made progress toward the achieving these goals.

- The Executive Committee's long-term goals involve working to solve the broad range of technical, procedural, and policy issues affecting UAS integration into the national airspace. In support of this objective, the Executive Committee agencies, other public agencies, and industry have also developed processes and procedures to safely demonstrate small UAS operations in remote areas of the Arctic, including beyond-line-of-site operations. The UAS demonstration occurred in domestic and international airspace on and off the coast of Alaska.

- RTCA Special Committee 228 has set out its own goals across two phases. Currently working toward completion of the first phase, RTCA is developing minimum operational performance standards for detect and avoid and command and control technologies for UASs. RTCA has made progress toward this goal with help from the Executive Committee. Specifically, the Executive Committee's Science and Research Panel developed a definition of "well clear" to help inform RTCA Special Committee 228's work.[29]
- The UAS Aviation Rulemaking Committee has a goal to develop a report for FAA on its efforts to provide direction for UAS operational criteria, among other tasks, by April 18, 2016.
- ASTM International F38's long-term goal involves developing and publishing voluntary consensus standards for small and large UASs as FAA requests them. ASTM International Committee F38 has developed standards and recommendations to support FAA's small UAS rulemaking that cover elements such as systems design, construction, and testing.

FAA has applied other interagency collaborative methods in support of UAS integration including memorandums of understanding or agreement (MOU)[30] and conferences. FAA entered into MOUs with DOD and NASA to expedite the COA process and ensure the availability of DOD's data.[31] According to FAA officials, the MOUs eased collaboration with DOD and NASA because the MOUs established roles and responsibilities for each agency as well as procedures for DOD to obtain COAs. In addition, FAA convenes meetings with test site officials and attends conferences where UAS issues are discussed. For example, FAA regularly holds conference calls and convenes technical interchange meetings with test site officials to address test site issues. According to FAA, the technical interchange meetings are opportunities for FAA to provide updates to the test sites and discuss common areas of research interest. The manager of the FAA's UAS Integration Office has presented information about its UAS efforts during industry conferences, such as the Association for Unmanned Vehicle Systems International's annual meeting. These conferences allow FAA to provide guidance and updates directly to the industry and public.

FAA's Designated Test Sites have Conducted UAS Test Flights and Addressed Challenges

FAA's Designated Test Sites are Operating and Conducting UAS Operations

Since being named in December 2013, the six designated test sites have become operational, applying for and receiving authorization from FAA to conduct test flights. Specifically, from April through August 2014, each of the six test sites became operational and signed an Other Transaction Agreement with FAA, establishing their research and development relationship. All flights at a test site must be authorized under the authority of a COA or an experimental certificate approved by FAA.[32] Since becoming operational, five of the six test sites received 48 COAs and one experimental certificate in support of UAS operations resulting in over 195 UAS flights across the five test sites. These flights provide the operations and safety data to FAA as required by the COA. While there are only a few contracts with industry thus far, according to test site operators, these will be important if the test sites are to generate sufficient revenue to remain in operation. Table 5 provides an overview of test-site activity since the sites became operational.

According to all test sites, FAA approval for access to the airspace can be a lengthy process taking 90 days or even longer. FAA and the test sites have found ways to allow quicker access to the test site airspace and relieve some administrative burden from FAA.

- In February 2015, FAA awarded the Northern Plains Test Site in North Dakota four broad area COAs that were aircraft specific. According to a test site official, these COAs allowed designated aircraft to fly over nearly the entire state of North Dakota and will make it easier to accommodate industry for research. Furthermore, these COAs were a positive step in allowing quicker access to the airspace at test sites. Reducing FAA's role in the process creates more certainty regarding how long it will take an operator to access airspace at the test sites. Specifically, the test site representative indicated FAA's role was reduced because there was a process that allowed aircraft to be added to these existing COAs that was simpler than applying for individual COAs.

Table 5. Overview of Activity at Five of Six Designated Unmanned Aerial Systems (UAS) Test Sites, as of March 2015

Type of test site activity	Overview since becoming operational in 2014
Total unmanned aerial systems (UAS) flights at FAA designated test sites	• Over 195 total UAS flights. • One test site has had over 80 UAS flights.
Number and types of certificates of waiver or authorizations (COA) received	• Five test sites hold 48 COAs. • One test site held 4 broad area COAs allowing flights over nearly the entire state by specific aircraft. • Four other test sites were seeking COAs for large flight ranges that could apply to any aircraft.
Number of special airworthiness certificates for experimental aircraft	• A certified representative reviewed and approved an aircraft to operate under an experimental certification at one test site. • Three test sites have representatives affiliated with the test site to review and approve aircraft for experimental certification.
Signed contracts with UAS companies	• Five test sites have 22 contracts with industry groups and companies to conduct UAS operations at their respective test site thus generating revenue for the test sites. • All test sites have additional negotiations with companies under way.

Source: FAA designated test sites. | GAO- 15-610

Note: Of FAA's six designated test sites, one— Texas A&M University-Corpus Christi test site—did not respond to our request for this information.

- In May 2015, FAA approved a "blanket" COA allowing the test sites to conduct UAS operations at or below 200 feet anywhere in the national airspace, similar to the authority provided to Section 333 exemptions. In particular, these COAs will be for small UAS operations, during the day, within line of sight of the operator, and the operations cannot occur in restricted airspace and areas close to airports. According to FAA, this will help improve UAS access allowing more operations in support of research that can further the UAS integration process. Previously, all UASs needed their own COA when operating at a test site but this action by FAA will allow any small UASs to operate at the test sites within the COA's requirements.

- The use of designated airworthiness representatives by the test sites to review and approve experimental certificates may be quicker for industry and relieve some of FAA's workload.[33] Industry benefits from not having to lease its aircraft to the test site, as all test sites are operated by public entities (academic institutions or federal, state, or local governments) and thus all aircraft must be public aircraft, unless they are operating under a special certificate.[34] In addition, any industry group working with the test site would not have to go through FAA to receive the experimental certificate. The Nevada test site has affiliated itself with a designated airworthiness representative, who has approved an aircraft to operate under an experimental certificate for the Nevada test site. According to FAA, the use of a designated airworthiness representative allows it to better leverage its resources.

FAA officials and some test site officials told us that progress has been made in part because of FAA's and test sites' efforts to work together. Test site officials meet every 2 weeks with FAA officials to discuss current issues, challenges, and progress. According to meeting minutes, these meetings have been used to discuss many issues from training for designated airworthiness representatives to processing of COAs. In addition, the six test sites have developed operational and safety processes that have been reviewed by FAA. Thus, while FAA has no funding directed to the test sites to specifically support research and development activities, FAA dedicates time and resources to supporting the test sites, and FAA staff we spoke to believe test sites are a benefit to the integration process and worth this investment.

Test Sites Faced Challenges during their First Year

Despite the progress made since they began operating, according to test site operators, they faced a number of challenges in the first year of operations:

- Guidance on research: According to FAA, because the test sites receive no federal funding, FAA can neither direct specific research to be conducted nor direct the test sites to share specific research data, other than the operations and safety data required by the COA. The Other Transaction Agreement for each test site defines the purpose of the test sites as a place to conduct research and testing under FAA safety oversight to support UAS integration into the national airspace. The Other Transaction Agreement indicates the test sites will provide FAA with UAS research and operational data to support the development of procedures, standards, and regulations. However, FAA officials told us that the Anti deficiency Act may prevent the agency from directing specific test site activities without providing compensation.[35] In October 2014, FAA provided a list of potential research areas to the test sites to guide the research that each test site may conduct. According to FAA, this document was not to be construed as a directive but more as guidance for possible research areas. However, three test sites told us this document was too broad to be considered guidance for the research test sites should conduct.

Companies Conducting Beyond - Visual-Line-of-Sight Testing Overseas

Insitu, Inc., a Boeing subsidiary, conducted beyond-visual-line-of-sight testing in Denmark in May, 2015 with a ScanEagle UAS. The flights took place in cooperation with the Danish Transport Authority as part of an agreement signed by Boeing and the airport to develop a UAS Test Center in Denmark, which is used for training, testing, and development. The activity included members of the public and private sector, including the UAS Denmark Consortium, a group of companies, government organizations, and other entities supporting UAS industry development. The testing demonstrated capabilities for a variety of industries, including agriculture and aerial surveying, emergency and natural-disaster response, and defense and Arctic surveillance.

[An Insitu, Inc. ScanEagle UAS launch at sea.]
Source: Insitu, Inc. | GAO-15-610

- Incentives for industry to use test sites: Five test sites told us that the UAS industry gains little advantage to using the test sites for research because they must follow similar processes to access the airspace anywhere. As previously indicated, all flights at a test site must be authorized under the authority of a COA or an experimental certificate, two means the industry can use to access the airspace anywhere, not just at the test sites. All six test sites are seeking to make it easier for companies to access the airspace at a test site rather than working external of the test sites. For example, all test sites told us a COA applicable to any aircraft and covering large flight areas would be beneficial. This would be similar to the two COAs previously discussed, the COA received by the Northern Plains test site and the "blanket" 200-foot COA recently authorized by FAA. However, representatives from two test sites indicated that the "blanket" 200- foot COA meets some of their needs as it is applicable only to public aircraft, meaning that civil operators still have to lease the aircraft to the test site for operations. But, as one test site representative stated a broad area COA, allowing civil operations at the test site would be even more beneficial. Another test site representative indicated that they will continue to work with FAA to make access easier allowing flights at higher altitude with different aircraft.
- Maintaining operations: While all the test sites had some level of initial funding, from either private industry or state legislatures, to become operational, they must attract UAS industry to the test sites to generate enough revenue to maintain operations. However, test site operators reported that test sites have additional requirements as opposed to operating outside the test sites, including leasing the aircraft to the test site to operate under the public COA. While test sites have signed 22 contracts, there is a chance that some test sites will not survive due to the financial burden. Some companies have made a decision to go to other countries to conduct UAS testing because they believe it takes less time to be approved for test flights. For example, Amazon has reported it has testing under way in multiple countries outside the United States, including a site in Canada. In an effort to attract some industry operators, the Pan Pacific Test Site has a location in Iceland where, according to the Director, review and approval for test flights can happen much faster, in as few as 10 days, relative to over 90 days a COA may take in this country.

In addition, the UAS industry is conducting tests in this country outside the test sites. For example, CNN worked with Georgia Institute of Technology.

FAA Supports Additional Research and Development Efforts

FAA has used cooperative research and development agreements, federally funded research and development centers, and grants to conduct other UAS research and development. These agreements for research are similar to the Other Transaction Agreement that directs the purpose and goals of the relationship between FAA and the research entities. However, unlike the Other Transaction Agreement in place for the test sites, according to FAA, many of these agreements have language specifically addressing the sharing of research and data. The following are examples of other resources FAA has devoted to UAS integration research and development:

- Cooperative research and development agreement: New Mexico State University has had a flight test center operating for several years under a cooperative research and development agreement with FAA. The center serves a similar purpose to the designated test sites but has been operating since 2007. The flight test center has conducted research in many areas including nighttime flying and more recently research into long-endurance UAS flights operating between 10,000 and 17,000 feet. According to an official, the New Mexico State University's flight test center has challenges with getting access to the airspace for customers because the process to receive a COA can be lengthy. In addition, this official told us the flight test center would like authority to approve COAs to operate at the test center because the FAA is backlogged and therefore approvals are delayed. Finally, according to the flight test center operators, FAA can get data from the research being conducted at the test site but does not direct them what to provide. While the flight test center has operated under a Cooperative Research and Development Agreement since 2007, in May 2015 the Flight Test Center switched to an Other Transaction Agreement to continue UAS testing.
- Federally funded research and development center: MITRE manages federal funded research and development centers for multiple federal agencies including FAA and DOD. MITRE has ongoing work

supporting FAA's UAS integration effort by supporting UAS standards and rulemaking and supporting research planning and progress, among other efforts. MITRE brings together the federal agencies— FAA, NASA, DOD, DHS, and others—to advance UAS integration. In its role, according to MITRE officials, one of the biggest challenges is how to integrate all the UAS-related work across the federal government, academia, and private sector.

- Grants: In August 2014, FAA awarded two grants to Georgia Tech Research Corporation and the University of North Dakota to conduct literature reviews of UAS issues. Georgia Tech is collecting information on research being conducted on the effect of UAS collisions on other airborne and ground based objects. The University of North Dakota is looking at the UAS safety criteria and particularly if UASs could be deadly. According to FAA, both studies will support ongoing UAS research and help determine the applicability of past studies.
- Center of Excellence: In May 2015, FAA selected a team led by Mississippi State University as the Center of Excellence for UAS.[36] According to FAA, the goal of the Center of Excellence will be to create a cost-sharing relationship among academia, industry, and government that will focus on the primary research areas needed to support UAS integration. FAA hopes the center could provide both short- and long-term research through testing and analysis. In support of it serving this purpose, the Center of Excellence has an annual $500,000 budget for the next 10 years.

FAA also has additional resources to support the UAS integration, including facilities working on research and development and management of FAA's other research and development efforts for UAS integration. FAA's William J. Hughes Technical Center houses staff in charge of supporting and managing FAA's designated test sites.[37] While the test sites do not have specific funding, FAA has dedicated resources located at Hughes Technical Center to support the set up and ongoing operations of the test sites. For example, COA data are collected and analyzed at the Hughes Technical Center. In addition, FAA has participated in the twice-a-year technical interchange meetings with the test sites. These meetings have brought together the test sites and FAA to address issues in the set-up and operation of the test sites. Furthermore, FAA has staff supporting the test sites through review of test site operation and safety procedures and manuals to support the monthly reporting of the operational and safety data required by each COA.

OTHER COUNTRIES HAVE ESTABLISHED REGULATIONS, AND SOME ALLOW MORE COMMERCIAL OPERATIONS THAN THE UNITED STATES

Numerous Countries Allow UAS Operations

According to numerous studies and stakeholders we interviewed, many countries around the world, have been allowing commercial UAS operations in their airspace for differing purposes. We also identified a number of countries that allow commercial UAS operations and have done so for years. Specifically, Canada and Australia have regulations pertaining to UAS that have been in place since 1996 and 2002, respectively. According to a recent MITRE study, the types of commercial operations allowed vary by country and include aerial surveying, photography, and other lines of business.[38] For example, Japan has allowed UAS operations in the agriculture industry since the 1980's to help apply fertilizer and pesticide.[39]

In March 2015, the European Aviation Safety Agency (EASA) issued a proposal for UAS regulations that creates three categories of UAS operations—open, specific, and certified.[40] This proposal seeks the safe integration of UAS into national airspace, as well as support for the UAS industry in Europe. Generally, the open category would not require authorization from an aviation authority but would have basic restrictions including altitude and distance from people. The specific category would require a risk assessment of the proposed operation and an approval to operate under restrictions specific to the operation. The final proposed category, certified operations, would be required for those higher-risk operations, specifically when the risk rises to a level comparable to manned operations. This category goes beyond FAA's proposed rules by proposing regulations for large UAS operations and operations beyond the pilot's visual line of sight. As other countries work toward integration, standards organizations from Europe and the United States are coordinating to try and ensure harmonized standards. Specifically, RTCA and the European Organization for Civil Aviation Equipment (EUROCAE) have joint committees focused on harmonization of UAS standards.[41]

UAS Regulations in Australia, Canada, France, and the United Kingdom are Similar to Each Other and to Proposed Regulations in this Country

We studied the UAS regulations of Australia, Canada, France, and the United Kingdom and found that these countries impose similar types of requirements and restrictions on commercial UAS operations. For example, all these countries except Canada require government-issued certification documents before UASs can operate commercially.[42] In addition, each country requires that UAS operators document how they ensure safety during flights and their UAS regulations go into significant detail on subjects such as remote pilot training and licensing requirements. For example, the United Kingdom has established "national qualified entities" that conduct assessments of operators and make recommendations to the Civil Aviation Authority as to whether to approve that operator. Similar regulations in these countries continue to evolve. In November 2014, Canada issued new rules creating exemptions for commercial use of small UASs weighing 4.4 pounds or less and from 4.4 pounds to 55 pounds. UASs in these categories can commercially operate without a government-issued certification but must still follow operational restrictions, such as a height restriction and a requirement to operate within line of sight. Transport Canada officials told us this arrangement allows them to use scarce resources to regulate situations of relatively high risk. Australia, in similar fashion, is considering relieving UASs lighter than 4.4 pounds from its requirement to obtain a UAS Operator's certificate. France's regulation describes 4 weight-based categories of UAS as well as 4 operational scenarios of increasing complexity. The regulations then discuss which UAS categories can operate in each scenario. FAA, by electing to focus on UASs up to 55 pounds in its Small UAS NPRM, has taken a similar risk-based approach.

Firefighting with UASs in Australia

Firefighters in Western Australia have used small UASs during their firefighting operations. The UASs provide intelligence on hot spots and containment lines at night when manned aircraft are not permitted to fly.

[An unmanned Lockheed Martin/Kaman K-MAX helicopter extinguishes several fires, aided by the small unmanned Indago Quad Rotor, also from Lockheed Martin, during a demonstration.]

[The Lockheed Martin Indago Quad Rotor.]

Source: Furnished with permission of Lockheed Martin Corporation.

The United States has not yet finalized regulations specifically addressing its small UAS operations, but if UASs were to begin flying today in the national airspace system under the provisions of FAA's proposed rules, their operating restrictions would be generally similar to regulations in these other four countries. However, there would be some differences in the details. For example, FAA proposes altitude restrictions of below 500 feet, while Australia, Canada, and the United Kingdom restrict operations to similar but slightly lower altitudes. Other proposed regulations require that FAA certify UAS pilots prior to commencing operations, while Canada and France do not require pilot certification in certain low risk scenarios. While FAA continues to finalize the small UAS rule—a process that could take until late 2016 or early 2017—other countries continue to move ahead with UAS integration. Thus, when the rule is finalized the operating restrictions in this country may be well behind what exists in other countries if the final rule reflects the proposed rule. Table 6 shows how FAA's proposed rules compare with the regulations of Australia, Canada, France, and the United Kingdom.

Table 6. Comparison of Regulatory Requirements for Commercial Unmanned Aerial Systems (UAS) Operations in Select Countries

Regulatory requirements for commercial Unmanned Aerial Systems (UAS)	United States (proposed)	Australia	Canada	France	United Kingdom
Weight classifications (in pounds)	< 55	≤ 0.2 0.2 < ≥ 331 > 331[a]	<4.4 4.4 <> 55 >55	<4.4 4.4 ≤> 55 55 ≤ > 331 ≥ 331	≤15 15 < ≥ 44 44 <> 331 ≥ 331
Government-issued documents for airspace access	Unmanned aircraft operator certificate	UAS operator certificate	None; meet specified conditions for < 55 lbs.[b]	Authorization	Permission from Civil Aviation Authority
Pilot training or certification required	Certification[c]	Certification	Training	Operator certifies pilot qualification[d]	Certification[e]
Pilot proficiency check	Biennially	Annually	Upon application	None[f]	None[g]
Airworthiness certification required	No	No, for < 331 lbs.	No, for < 55 lbs.	No < 55 lbs.	Depends upon the weight of the UAS[h]
Beyond line of sight operations allowed by regulation	No	Not specifically addressed in regulation[i]	No	No[j]	No[k]
Restrictions from congested or built-up area	May operate over congested area but not over any persons not involved with the operations.	Unless specifically authorized must fly at sufficient height where, if any of its components fail, it would be able to clear the area.	5.75 miles	Do not allow flying over these areas without special permission	Limited, based on case-by-case review
Altitude restrictions	500 ft.	Unless specifically authorized, 400 ft.	300 ft. if < 55 lbs.	492 ft.	400 ft.

Source: GAO analysis of UAS regulations in foreign countries. | GAO-15-610

Note: In certain instances, UASs may operate outside the requirements shown in this table with prior approval from the cognizant authority, e.g., the civil aviation authority or air traffic control. This table shows regulatory requirements for the UAS category used most frequently in each country.

[a] Regulations for large fixed-wing UASs (exceeding 331 pounds) apply to helicopters or rotary wing UASs heavier than 221 pounds.

[b] Operators must have certain documents available, including the exemption regulation and proof of liability insurance.

[c] Pilot must pass a test at an FAA-approved testing center.

[d] Owner of the UAS must self-certify that the UAS pilots have the required training.

[e] Pilot qualifications are determined on a case-by-case basis based a number of factors including pilot experience and aircraft weight. Permission is not required for aircraft 44 pounds or less being flown within direct unaided line of sight and away from people, property, and congested areas.

[f] Owner of the UAS must self-certify that the UAS pilots have the required training if the pilot has not flown the UAS in the preceding 12 months.

[g] Self-certification accepted using logbook entries unless the pilot changes aircraft type, or the pilot has less than 2-hours experience logged in preceding 3 months on same aircraft type.

[h] An airworthiness certification is not required for UAS weighing less than 44 pounds but is required for UAS weighing more than 44 pounds Exemptions may be available in specified circumstances.

[i] An Advisory Circular issued by Australia's Civil Aviation Safety Authority (AC 101-1(0)) provides for beyond line of sight operations under certain circumstances.

[j] UAS operations beyond line of sight are very limited and allowed only on a case-by-case basis, according to a French civil aviation official.

[k] Beyond line of sight operations are allowed with aircraft fitted with a detect-and-avoid system or, operated within a segregated airspace. The Civil Aviation Authority has noted that it is not aware of any detect-and-avoid system with adequate performance and reliability, but has several areas of segregated airspace.

While regulations in these countries generally require that UAS operations remain within the pilot's visual line of sight, some countries are moving toward allowing limited operations beyond the pilot's visual line of sight. For example, according to Australian civil aviation officials, they are developing a new UAS regulation that would allow operators to request a certificate allowing beyond-line-of-sight operations. However, use would be very limited and allowed only on a case-by-case basis. Similarly, according to a French civil aviation official, France approves on a case-by-case basis, very limited beyond-line-of-sight operations. Finally, in the United States, there have been beyond-line-of-sight operations in the Arctic, and NASA, FAA and the UAS industry have successfully demonstrated detect-and-avoid technology, which is necessary for beyond line-of-sight operations.

Like the United States, Australia, Canada, France, and the United Kingdom distinguish between recreational model aircraft and commercial UASs and have issued guidelines for safe operation.[43] For example, the United Kingdom defines model aircraft as any small unmanned aircraft, weighing less than 44 pounds, or large unmanned aircraft weighing more than 44 pounds, that is used for sporting and recreational purposes. Australia makes no practicable distinction between a small UAV and a model aircraft except that of use—model aircraft are flown only for the sport of flying them. However, Australia also defines a giant model aircraft as one weighing between 55 pounds and 331 pounds.

Approvals for commercial UAS operations have increased in these four countries and some allow more commercial operations. Since 2011 the number of approvals for commercial operations in France has increased every year. According to a Civil Aviation Authority official, in 2014, there were about 3,600 commercial UAS operators. In Canada, according to a Transport Canada official, there were over 1,600 approvals for commercial and research related UAS operations in 2014. As previously mentioned, certain commercial operations in Canada do not need approval as of November 2014 and there may be even more UAS operations. The United Kingdom's Civil Aviation Authority attributes the growth to UASs in their country to the UASs' becoming less expensive and simpler to operate. In the United Kingdom, as of February 2015, there were 483 commercial UAS operators, and this number has increased every year since 2010. Similar to the United Kingdom, Australia has seen an increase in commercial UAS operators since 2010 with currently over 200 approved commercial operators. Australia's Parliament attributes the growth to improvements in UASs' piloting and control technologies, as well as reductions in UAS prices. With FAA's approvals for commercial exemptions exceeding 500 as of June 9, 2015, the United States has closed the gap with some other countries' level of commercial use.

Other Countries Face Common Challenges to UAS Integration

Other countries face challenges that are common across some countries, including the United States, trying to integrate UAS operations. Specifically, some of the challenges are:

- Technology shortfalls and unresolved spectrum issues. Technology needs and concerns about available spectrum constrain full integration of UASs into airspace with manned aircraft in the United States and in countries around the world. UASs' current inability to detect and avoid other aircraft, the lack of a standard for command and control systems, and no dedicated and secure frequency spectrum are technical challenges preventing full UAS integration into the national airspace. However, organizations around the world are looking to address these technology issues and develop standards to support safe UAS operations. At the worldwide level, the International Civil Aviation Organization is addressing how UAS integration would affect its existing standards. At the European level, the European UAS Roadmap contains a strategic research and development plan that describes anticipated deliverables along with key milestones, timelines, and resources needed. Separate from the international organizations, researchers in individual countries are also addressing these challenges. For example, in February 2014, Australian researchers achieved what was then believed to be a world-first breakthrough for small UASs by developing an onboard system that has enabled a UAS to detect another aircraft using vision while in flight.
- Safe operations by recreational users. Countries around the world also face challenges in ensuring that UAS purchasers operate them safely. As UASs become more affordable, and increasingly available some individuals are conducting unsafe or illegal UAS operations. In July 2014, Australia's Parliament reported on testimony, from several witnesses, that UASs are being flown by operators who unknowingly break safety rules, thereby posing a safety risk to manned aircraft and persons on the ground. In response to unsafe operations, a few

countries have placed outright bans on UAS operations. For example, in India, in response to the surge in interest for commercial and recreational use, the government placed an outright ban on any UAS use until its civil aviation agency issues regulations. Similar to the "Know Before You Fly" education campaign in this country, other countries have sought to educate operators. For example, the United Kingdom has developed and distributed a brochure describing safe flying practices. In Australia, UAS purchasers receive a similar document when they purchase the product. Canada has launched a national safety awareness campaign for UASs, which aims to help Canadians better understand the risks and responsibilities of flying UASs. In addition, Transport Canada has set up a web page that provides safe guidelines for flying UASs and answers frequently asked questions

While countries face some UAS integration challenges that are similar to the United States, other challenges such as airspace complexity and ease of regulatory change, can make integration in this country more difficult. Airspace complexity is one aspect in which the United States differs from other countries. According to FAA, the U.S. airspace is the busiest and most complex in the world, where UASs, after integration, would share with more than 300,000 general aviation aircraft, ranging from amateur-built aircraft, rotorcraft, and balloons, to highly sophisticated turbojets (executive jets).[44] Introducing potentially large numbers of UAS by hobbyists, farmers, law enforcement agencies, and others would add to this complexity. In contrast, according to a study by MITRE, other countries have fewer aviation aircraft, a situation that may make integrating UAS easier. For example, the U.K. has about 20,000 registered general aviation aircraft, while Australia has around 8,400. A study conducted by MITRE for FAA indicated this factor as one that can affect the speed of change and adaptation in various aviation environments.

AGENCY COMMENTS

We provided a draft of this report to Department of Transportation (DOT) for review and comment. In comments, which were provided in an email, DOT stated that the report addresses many of the challenges of UAS integration but does not address any environmental concerns and that the report should state that it did not examine the environmental considerations of UAS integration. DOT further noted that FAA is conducting research to understand the environmental impacts of UAS integration, the role that UASs play in National Environmental Policy Act compliance, and the applicability of noise standards regulations to UASs.[45] We did not examine the environmental considerations of UAS integration. The discussion of challenges in the report does not mention environmental concerns because it focuses on challenges the test sites faced during their first year of operation, as reported by the test site operators. We did clarify in our scope and methodology description that we did not cover environmental considerations of UAS integration. DOT also provided technical comments on the draft that we incorporated as appropriate.

Gerald L. Dillingham, Ph.D.
Director, Physical Infrastructure Issues

APPENDIX I: OBJECTIVES, SCOPE, AND METHODOLOGY

This report focuses on FAA's efforts to develop procedures to allow UAS use within the national airspace system. Specifically, we reviewed (1) the status of FAA's progress toward safe integration of UAS into the national airspace, (2) research and development support from FAA's test sites and other resources, and (3) how other countries have progressed toward UAS integration into their airspace for commercial purposes.

To address the three objectives, we reviewed and synthesized a range of published reports from GAO[46] and FAA that included general background information on a variety of related issues, such as FAA's framework for UAS integration, efforts to accommodate ongoing research and commercial UAS use, and UAS technology challenges.[47] We reviewed other relevant background literature on related issues, including results from databases, such as ProQuest® and Nexis®, trade publications, literature from industry stakeholder groups, and information from the Internet. We also reviewed provisions of the FAA Modernization and Reform Act of 2012,[48] and the Notice of Proposed Rulemaking for small UAS operations.[49] In addition, we reviewed more detailed and specific documentation related to the different objectives, as described below.

To determine FAA's progress toward safe integration of UAS into the national airspace, we:

- Reviewed documents provided by officials and conducted semi-structured interviews with officials at federal agencies, including the FAA's Unmanned Aircraft Systems Integration and Research and Development Offices, the Department of Defense (DOD), the National Aeronautics and Space Administration (NASA), and the Department of Homeland Security. We reviewed FAA's Comprehensive Plan[50] and Roadmap for UAS integration.[51]
- Interviewed representatives from FAA's Joint Planning and Development Office, the UAS Aviation Rulemaking Committee, RTCA, and MITRE Corporation as well as voluntary standards development organization ASTM International. We interviewed representatives from the Association for Unmanned Vehicle Systems International, Aircraft Owners and Pilots Association, American Institute of Aeronautics and Astronautics, and the Academy of Model Aeronautics. We also obtained information on the Federal Modernization and Reform Act of 2012 Section 333 exemptions FAA granted from FAA and http://www.regulations.gov[52] from September 2014 to May 2015.
- Reviewed documents provided by and interviewed federal and industry representatives from the collaborative groups—the Executive Committee, RTCA Special Committee 228, UAS Aviation Rulemaking Committee, and ASTM International Committee F38—and industry groups that are involved in FAA's efforts to integrate UAS into the national airspace system

To identify research and development support from FAA's test sites and other resources, we:

- Reviewed and analyzed documents from each of the six test sites where FAA has recently allowed UAS operations including the applications submitted by the selected test sites and quarterly reports provided to FAA.
- Conducted semi-structured interviews with officials from the test sites, including the State of Nevada, the University of Alaska, the North Dakota Department of Commerce, Griffiss International Airport, the Virginia Polytechnic Institute & State University, and Texas A&M University Corpus Christie to determine the issues encountered in an effort to become operational, conduct research,

share the research results with FAA, and receive support or guidance from FAA.

- Spoke with representatives from other universities with centers of research on UAS technology and issues, including New Mexico State University, Massachusetts Institute of Technology Lincoln Laboratory, the Humans and Autonomy Lab at Duke University, and the Georgia Institute of Technology to obtain information about the resources FAA has dedicated to conducting other UAS research and development.

To identify how the United States compares to other countries in the progress and development of UAS use for commercial purposes, we:

- Developed case studies for four countries that have made progress in integrating UASs into their national airspace—France (Direction générale de l'aviation civile); the United Kingdom (UK Civil Aviation Authority); Australia (Australia Civil Aviation Safety Authority); and Canada (Transport Canada Civil Aviation). We selected these countries based on several factors including the status of regulatory requirements for commercial UASs, beyond-line-of-site activities, and whether the country allows non-military UAS to operate in the airspace. We obtained the UAS regulations of each country and interviewed civil aviation authorities in each to obtain additional information about the issues encountered with UAS.
- Interviewed other stakeholders familiar with the UAS activities currently occurring in other countries to determine the factors that influenced their country's policies regarding UASs including the International Civil Aviation Organization (ICAO).[53]

We conducted this performance audit from January 2014 to July 2015 in accordance with generally accepted government auditing standards. Those standards require that we plan and perform the audit to obtain sufficient, appropriate evidence to provide a reasonable basis for our findings and conclusions based on our audit objectives. We believe that the evidence obtained provides a reasonable basis for our findings and conclusions based on our audit objectives.

Appendix II: Selected Requirements and Status for Unmanned Aerial Systems (UAS) Integration under the Federal Aviation Administration (FAA) Modernization and Reform Act of 2012, as of December 2014

Deadline	FAA Modernization and Reform Act of 2012 requirement	Status of action
05/14/2012	Enter into agreements with appropriate government agencies to simplify the process for issuing Certificates of Waiver or Authorization (COA) or waivers for public unmanned aerial systems (UAS).	In process—memorandum of agreement (MOA) with DOD signed Sept. 2013; MOA with Department of Justice signed Mar. 2013; MOA with NASA signed Mar.2013; MOA with Department of Interior signed Jan. 2014; MOA with the Office of the Director, Operational Test and Evaluation (DOD) signed Mar. 2014; MOA with National Oceanic and Atmospheric Administration still in draft.
05/14/2012	Expedite the issuance of COAs for public safety entities	Completed
08/12/2012	Establish a program to integrate UASs into the national airspace at six test ranges. This program is to terminate 5 years after date of enactment.	Completed
08/12/2012	Develop an Arctic UAS operation plan and initiate a process to work with relevant federal agencies and national and international communities to designate permanent areas in the Arctic where small unmanned aircraft may operate 24 hours per day for research and commercial purposes.	Completed

(Continued)

Deadline	FAA Modernization and Reform Act of 2012 requirement	Status of action
08/12/2012	Determine whether certain UAS can fly safely in the national airspace before the completion of the Act's requirements for a comprehensive plan and rulemaking to safely accelerate the integration of civil UASs into the national airspace or the Act's requirement for issuance of guidance regarding the operation of public UASs including operating a UAS with a COA or waiver.	Completed
11/10/2012	Develop a comprehensive plan to safely accelerate integration of civil UASs into national airspace.	Completed
11/10/2012	Issue guidance regarding operation of civil UAS to expedite COA process; provide a collaborative process with public agencies to allow an incremental expansion of access into the national airspace as technology matures and the necessary safety analysis and data become available, until standards are completed and technology issues are resolved; facilitate capability of public entities to develop and use test ranges; provide guidance on public entities' responsibility for operation.	Completed
02/12/2013	Make operational at least one project at a test range.	Completed

Deadline	FAA Modernization and Reform Act of 2012 requirement	Status of action
02/14/2013	Approve and make publically available a 5-year road map for the introduction of civil UAS into national airspace, to be updated annually.	Completed
02/14/2013	Submit to Congress a copy of the comprehensive plan.	Completed
08/14/2014	Publish in the Federal Register the Final Rule on small UAS.	In process
08/14/2014	Publish in the Federal Register a Notice of Proposed Rulemaking to implement recommendations of the comprehensive plan.	None to date
08/14/2014	Publish in the Federal Register an update to the Administration's policy statement on UAS in Docket No. FAA-2006-25714.	None to date
09/30/2015	Achieve safe integration of civil UAS into the national airspace.	In process
12/14/2015	Publish in the Federal Register a Final Rule to implement the recommendations of the comprehensive plan.	None to date
12/31/2015	Develop and implement operational and certification requirements for public UAS in national airspace.	In process
05/14/2017	Report to Congress on the test ranges	None to date

Source: GAO analysis of FAA information. | GAO-15-610

End Notes

[1] See GAO, *Unmanned Aerial Systems: Department of Homeland Security's Review of U.S. Customs and Border Protection's Use and Compliance with Privacy and Civil Liberty Laws and Standards*, GAO-14-849R (Washington, DC: Sept. 30, 2014). UAS aircrafts do not carry a pilot onboard, but instead operate by following commands from pilot-operated ground control stations or pre-programmed routes. UASs are also referred to as "unmanned aerial vehicles," "unmanned aircraft systems," "remotely piloted aircraft," "unmanned aircraft," or "drones." The term "unmanned aerial system" is used to recognize that UASs include not only the airframe and its power source, such as a battery or combustible engine, but also associated elements such as a ground control station and communications links.

[2] GAO, *Unmanned Aircraft Systems: Continued Coordination, Operational Data, and Performance Standards Needed to Guide Research and Development*, GAO-13-346T (Washington, DC: Feb. 15, 2013); *Unmanned Aircraft Systems: Measuring Progress and Addressing Potential Privacy Concerns Would Facilitate Integration into the National Airspace System*, GAO-12-981 (Washington, DC: Sept. 14, 2012); and *FAA Reauthorization Act: Progress and Challenges Implementing Various Provisions of the 2012 Act*, GAO-14-285T (Washington, DC: Feb. 5, 2014).

[3] We did not review the environmental challenges of UAS integration.

[4] Pub. L. No. 112-95, 126 Stat. 11.

[5] 80 Fed. Reg. 9544 (Feb. 23, 2015).

[6] RTCA was formerly known as the Radio Technical Commission for Aeronautics. ASTM International was formerly known as the American Society for Testing and Materials.

[7] ICAO is the international body that, among other things, promulgates international standards and recommended practices in an effort to harmonize global aviation standards.

[8] FAA's Federally Funded Research and Development Centers are located at MITRE, MIT's Lincoln Laboratory, and the Air Force Research Laboratory.

[9] "Detect and avoid" is needed to provide UAS with the capability to operate near other aircrafts and objects, yet avoid collisions. "Command and control" addressed the ability to maintain the integrity of radio signals to ensure that the UAS operates as expected and as intended. Dedicated radio-frequency spectrum is important because spectrum is a key security and safety vulnerability and because intentional or unintentional interruption of radio transmissions can sever the UAS's only means of control. The study of the "human factors" affecting UAS operators is important because the separation of pilot and aircraft 1) removes sensory cues needed for flight control, and 2) creates difficulties in scanning the visual environment surrounding the unmanned aircraft.

[10] Other Transaction Agreements are administrative vehicles used by a federal agency. Other Transaction Agreements generally enable the federal government and others entering into these agreements to freely negotiate provisions that are mutually agreeable. FAA's Other Transaction Agreement authority, at 49 U.S.C. § 106(l)(6) provides, in part, that the Administrator may enter into other transactions with any federal agency, among others, on such terms and conditions as the Administrator may consider appropriate.

[11] Pub. L. No. 112-95, §§ 331 – 335, 126 Stat. 11.

[12] The test sites are located at the University of Alaska (includes test ranges in Hawaii, Oregon, and Iceland); State of Nevada; New York's Griffiss International Airport (includes test range locations in Massachusetts); North Dakota Department of Commerce; Texas A&M University–Corpus Christi; and Virginia Polytechnic Institute and State University (Virginia Tech) (includes test ranges in Maryland, partnered with the University of Maryland, and New Jersey, partnered with Rutgers University).

[13] GAO, *Unmanned Aerial Systems: Efforts Made toward Integration into the National Airspace Continue, but Many Actions Still Required*, GAO-15-254T (Washington, DC: Dec. 10, 2014).

[14] The integration or use of military aircraft was outside the scope of this review.

[15] Joint Planning and Development Office, *Unmanned Aircraft Systems (UAS) Comprehensive Plan: A Report on the Nation's UAS Path Forward* (Washington, D.C.: Sept., 2013).

[16] The six goals address small UAS (under 55 pounds) operating within visual line-of-sight, larger UASs and operations beyond visual line-of-sight, planning and managing growing automation capabilities through research, and the opportunity for the U.S. to remain world leaders in UAS technology.

[17] DOD, *Unmanned Systems Integrated Roadmap FY2013-2038* (Washington, D.C.: 2013).

[18] FAA, *Integration of Civil Unmanned Aircraft Systems (UAS) in the National Airspace System (NAS) Roadmap: First Edition—2013* (Washington, D.C.: Nov. 2013).

[19] Department of Transportation Office of Inspector General, *FAA Faces Significant Barriers to Safely Integrate Unmanned Aircraft Systems into the National Airspace System*, Report Number AV-2014-061 (Washington, D.C.: June 2014).

[20] Deborah Kirkman, *Inputs and Recommendations for the Federal Aviation Administration UAS Integration Strategy: 2014–2019*, MITRE Corporation (August: 2014).

[21] FAA told us they do intend to publicly issue the implementation plan.

[22] 80 Fed. Reg. 9544 (Feb. 23, 2015).

[23] See Pub. L. No. 112-95, §336(c), 126 Stat. 11, 77.

[24] FAA's *Advisory Circular 91-57* sets out model aircraft operating standards that encourage voluntary compliance with specified safety standards for model aircraft operators. In June 2014, FAA also clarified this guidance in its Notice of Interpretation of the 2012 Act's Special Rule for Model Aircraft. 79 Fed. Reg. 36172 (June 25, 2014). The Academy of Model Aeronautics has also published voluntary guidance documents for its members.

[25] More specifically, section 336(a) of the 2012 Act prohibits the FAA from promulgating rules regarding model aircraft that meet all of the following statutory criteria: The aircraft is flown strictly for hobby or recreational use; The aircraft is operated in accordance with a community-based set of safety guidelines and within the programming of a nationwide community-based organization; The aircraft is limited to not more than 55 pounds unless otherwise certified through a design, construction, inspection, flight test, and operational safety program administered by a community-based organization; The aircraft is operated in a manner that does not interfere with and gives way to any manned aircraft; and when flown within 5 miles of an airport, the operator of the aircraft provides the airport operator and the airport air traffic control tower (when an air traffic facility is located at the airport) with prior notice of the operation.

[26] Section 336(b) of the 2012 Act provides that "[n]othing in this section shall be construed to limit the authority of the Administrator to pursue enforcement action against persons operating model aircraft who endanger the safety of the national airspace system. The proposed rule provides that it is FAA's existing authority, under 49 U.S.C. §§ 40103(b) and 44701(a)(5) that provide the FAA with the power to pursue enforcement "against persons operating model aircraft who endanger the safety of the national airspace system."

[27] Previously, an operator had to apply for and receive a COA for a particular block of airspace, a process that can take 60 days.

[28] Know Before You Fly (www.knowbeforeyoufly.org) was founded by three organizations with a stake in UAS safety: the Association for Unmanned Vehicle Systems International, the Academy of Model Aeronautics, and the Small UAV Coalition. The FAA is partnering with the founding members to spread the word about safe and responsible flying. FAA also stated that three manufacturers agreed to voluntarily include Know Before You Fly information for model aircraft operators in their packaging.

[29] The Federal Aviation Regulations generally require that pilots remain "well clear" of other aircraft when flying in the national airspace system. 14 C.F.R. §§ 91.115, 91.181. RTCA will use a definition of well clear provided by the UAS Executive Committee's Science and Research Panel as a baseline for detect and avoid standards.

[30] A memorandum of understanding is a written agreement between more than one federal agency or department that defines the roles, responsibilities, and how each party will act for specific aspects of the relationship between the two participants.

[31] FAA also has UAS related memorandums of agreement or understanding with the U.S. Department of Justice, the U.S. Department of the Interior, the National Oceanic and Atmospheric Administration within the U.S. Department of Commerce, and the Academy of Model Aeronautics.

[32] All test sites are operated by public entities (academic institutions or federal, state, or local governments) and thus all aircraft must be public aircraft—unless they are operating under a special certificate.

[33] Designated airworthiness representatives are individuals acting as a representative of FAA trained to examine, inspect, test and issue aircraft certificates.

[34] For a private company to operate under a public COA at a test site, it must lease the aircraft to the public entity. A concern of some private companies has been the protection of proprietary or sensitive data when they lease the aircraft to the public operator of the test site. The experimental certificate eliminates this concern of protecting proprietary technology.

[35] The Antideficiency Act prohibits federal officers and employees from, among other things, accepting voluntary services except for emergencies involving the safety of human life or the protection of property. 31 U.S.C. § 1342.

[36] In addition to Mississippi State University, the other team members include: Drexel University; Embry Riddle Aeronautical University; Kansas State University; Kansas University; Montana State University; New Mexico State University; North Carolina State University; Oregon State University; University of Alabama, Huntsville; University of Alaska, Fairbanks; University of North Dakota; and Wichita State University.

[37] The William J. Hughes Technical Center is located in Atlantic City, New Jersey, and contains laboratories supporting aviation research, development, testing, and evaluation of air traffic control and aircraft safety among other aviation areas. It also serves as the primary facility supporting the Next Generation Air Traffic System.

[38] MITRE Corporation, *UAS International Harmonization: A Comparative Policy Assessment of Selected Countries*, Outcome 6, Output 4 (fiscal year 2014).

[39] According to the MITRE study, Japan's regulations also allow UAS operations for agricultural purposes with UASs weighing less than 220 pounds.

[40] EASA is the European Union Authority in aviation safety. The main activities of the organization include the strategy and safety management, the certification of aviation products and the oversight of approved organizations and EU Member States.

[41] RTCA is a private non-profit organization consisting of industry experts. RTCA provides a venue for public-private collaboration supporting consensus building on aviation modernization issues. EUROCAE is a non-profit organization dedicated to aviation standards. The organization is composed of members, who are specialized in technical fields of aeronautics.

[42] UASs lighter than 55 pounds are exempt from Canada's requirement to obtain a Special Flight Operations Certificate. Those heavier than 55 pounds or otherwise not complying with the exemption requirements must obtain a Special Flight Operations Certificate.

[43] As discussed earlier, for the United States, the model aircraft definition is in the 2012 Act. Rules for flying model aircraft are advisory, and compliance is voluntary. However, FAA has enforcement authority over persons operating model aircraft who endanger the safety of the national airspace system.

[44] General aviation includes all civil aviation operations other than scheduled air services and non-scheduled air transport operations for remuneration or hire.

[45] 14 CFR Part 36.

[46] GAO, *Unmanned Aircraft Systems: Continued Coordination, Operational Data, and Performance Standards Needed to Guide Research and Development*, GAO-13-346T

(Washington, DC: Feb. 15, 2013); *Unmanned Aircraft Systems: Measuring Progress and Addressing Potential Privacy Concerns Would Facilitate Integration into the National Airspace System*, GAO-12-981 (Washington, DC: Sept. 14, 2012); and *FAA Reauthorization Act: Progress and Challenges Implementing Various Provisions of the 2012 Act*, GAO-14-285T (Washington, DC: Feb. 5, 2014).

[47] We did not review the environmental challenges of UAS integration.

[48] Pub. L. No. 112-95, 126 Stat. 11.

[49] 80 Fed. Reg. 9544 (Feb. 23, 2015).

[50] Joint Planning and Development Office, *Unmanned Aircraft Systems Comprehensive Plan: A Report on the Nation's UAS Path Forward* (Washington, D.C.: September 2013).

[51] FAA, *Integration of Civil Unmanned Aircraft Systems in the National Airspace System Roadmap: First Edition—2013* (Washington, D.C.: November 2013).

[52] Regulations.gov is a website that allows the pubic to access and submit comments on agencies' regulatory documents published in the Federal Register.

[53] ICAO is the international body that, among other things, promulgates international standards and recommended practices in an effort to harmonize global aviation standards.

In: Unmanned Aerial Systems
Editor: Wilfred Sparks

ISBN: 978-1-63484-652-3

Chapter 2

STATEMENT OF MARGARET GILLIGAN, ASSOCIATE ADMINISTRATOR FOR AVIATION SAFETY, FEDERAL AVIATION ADMINISTRATION. HEARING ON "UNMANNED AIRCRAFT SYSTEMS: KEY CONSIDERATIONS REGARDING SAFETY, INNOVATION, ECONOMIC IMPACT, AND PRIVACY"*

Senator Ayotte, Senator Cantwell, Members of the Subcommittee:

Thank you for the opportunity to appear before you today to discuss the status of the safe integration of Unmanned Aircraft Systems (UAS) into the National Airspace System (NAS).

The FAA is safely and steadily integrating UAS into the largest, most complex aviation system in the world. At the same time, UAS technologies continue to advance at a rapid pace. Consequently, novel applications emerge challenging us to develop a regulatory framework that will allow for continued innovation while ensuring the safety of other users of the airspace and people and property on the ground.

* This is an edited, reformatted and augmented version of a statement presented March 24, 2015 before the Senate Committee on Commerce, Science & Transportation, Subcommittee on Aviation Operations, Safety, and Security.

The FAA Modernization and Reform Act of 2012 (2012 Act) established the framework for the integration of UAS into the NAS and tasked the FAA with the safe integration of civil UAS into the system by October 2015. We have followed through with Congress' intent in the 2012 Act and have completed milestones forming the foundation for future integration. This includes long-term planning for the future of integration, collaborative research and development with interagency partners and with industry, and the establishment of test sites and airspace for UAS research and development and testing.

Consistent with the authority in section 333 of the 2012 Act, the FAA has issued 48 exemptions that allow for commercial activity in the NAS in low-risk, controlled environments. An exemption may be granted after a two-step process. First, the Secretary of Transportation determines that, based on criteria set forth in the statute, the UAS does not pose a risk to those operating in the NAS, the general public, or national security and it can be safely operated without an airworthiness certificate. The FAA will then use its exemption authority to grant relief from FAA regulations that may apply. The exemption process allows the FAA to evaluate each request to determine what conditions are required to ensure that the operation will not create an adverse impact on safety. Once an exemption is granted, the applicant must then apply for a civil Certificate of Waiver or Authorization (COA), an FAA authorization issued by the Air Traffic Organization permitting the operator to use specific airspace to conduct the proposed operation.

Last month, we proposed a rule that would allow small unmanned aircraft systems to operate for commercial purposes without first obtaining an airworthiness certificate, section 333 exemption, or a COA. The proposal would cover many potential small UAS operations and would offer a flexible framework for the safe use of small unmanned aircraft, while accommodating future innovation in the industry. As proposed, the United States would have one of the most flexible UAS regulatory frameworks in the world.

In addition to near-term challenges, the FAA is looking ahead at what is next, and how to coordinate near and long-goals while leveraging available resources to address the most pressing risks to the system.

LAYING A FOUNDATION AND TAKING THE NEXT STEP FOR SAFE INTEGRATION

From the outset, we have worked closely and successfully with government partners and industry stakeholders to achieve milestones put forward by the Act. We developed two long-term planning documents, the Comprehensive Plan and a five-year Roadmap, in coordination with other governmental agencies and industry to safely accelerate the integration of civil unmanned aircraft systems in the NAS. We have worked with members of the UAS Executive Committee (ExCom) to leverage our collective assets and conduct research and development to overcome some of the largest barriers to UAS integration and ensure the continued safety of the NAS. The FAA has collaborated with the National Aeronautics and Space Administration (NASA) on studies advancing air traffic control interoperability with the future UAS use of detect-and-avoid (DAA) systems in controlled airspace. We continue to collaborate with members of industry on flight tests to validate RTCA[1] standards for DAA systems as well as command and control radios. RTCA began work on the standards at the request of the FAA in 2013 and they are scheduled for completion in 2016. These standards will help to resolve two of the difficult challenges facing the industry for integration of UAS into the NAS. NASA, the FAA, and industry partners have successfully demonstrated a proof-of-concept airborne DAA system and prototype radios for use as command and control systems for UAS.

In November 2012, the FAA released its Arctic Implementation Plan to establish permanent operational areas and corridor routes in the Arctic for the operation of small UAS as required by the 2012 Act. In July 2013, a restricted category type certificate was issued to Insitu's ScanEagle X200 and to AeroVironment's PUMA so that each UAS could conduct Arctic flights for commercial purposes. In September 2013, ConocoPhillips began using Insitu's ScanEagle for its marine mammal and ice surveys. In June 2014, BP began using AeroVironment's Puma AE to survey its pipelines, roads, and equipment at Prudhoe Bay, Alaska. Safety and operational data from the operators will be used to develop UAS operations and performance standards. The FAA has also issued 176 special airworthiness certificates in the experimental category for civil UAS, 34 of which are currently active. Special airworthiness certificates are issued for research and development, crew training, and market surveys.

In December 2013, the FAA selected six test sites for non-federal entities to test UAS technology and operations. By September 2014, all of the UAS test sites, which were selected based on geographic and climatic diversity, were operational and will help us gather operational data to foster further integration. Flights of unmanned aircraft have already been conducted at test sites, including flights for research on agricultural and wildlife monitoring and on law enforcement and emergency services support.

Once the test sites were ready, the FAA gave them priority for their first COA. The test sites all qualified as public entities so their initial operations were under the public aircraft operations statute. In 2014, the FAA implemented a Designated Airworthiness Representatives program which will permit test site designees to issue experimental certificates for unmanned aircraft for research and development, crew training, and market surveys. Test site designees need only complete FAA training, available online or in person, to be authorized to work within this new program. This new delegation authority will improve access to the test sites by UAS manufacturers, as well as help to decrease the workload on the FAA to process UAS experimental certificates.

On February 15, 2015, the FAA announced the Small UAS Notice of Proposed Rulemaking that would allow routine use of certain small UAS in the NAS. The proposed rule would allow unmanned aircraft weighing up to 55 pounds to operate without the need for an airworthiness certificate if the operations take place under a set of parameters to maintain safety including operating at speeds below 100 mph and below 500 feet in altitude. The proposal would allow operations during daylight hours and would require the operator to be able to see the unmanned aircraft at all times. Rather than requiring a private pilot certificate, the proposal is for operators to obtain a FAA unmanned aircraft operator's certificate by passing a written proficiency test. Before each flight, operators would conduct a preflight inspection, just as pilots do with manned aircraft today. The proposal does not permit flight over any persons not directly involved in the operation unless those persons are located under a covered structure. Also, unmanned flights would not be allowed in Class A (18,000 feet & above) airspace and, unless air traffic control gives permission, would be restricted from operating in certain busy airspace or in airspace otherwise restricted to most or all aviation users.

In April 2008, the FAA chartered the small UAS Aviation Rulemaking Committee (ARC) that included members from a wide spectrum across the aviation community, to provide recommendations on how small UAS could be safely integrated into the NAS. In April 2009, the small UAS ARC provided recommendations and the FAA began working on a rulemaking that encompassed the widest possible range of small UAS operations. The approach utilized a regulatory structure similar to the one used for manned aircraft; small UAS operations that pose a low risk to people, property, and other aircraft would be subject to less stringent regulation, while small UAS operations posing a greater risk would be subject to more stringent regulation to mitigate the greater risk. Developing this broadly-scoped approach to the rulemaking effort took significantly longer than anticipated, as the FAA had a desire to put forth a proposal that struck the right balance between mitigating safety risks, yet allowing for changing technology and innovation.

The framework for UAS integration established by the FAA Modernization and Reform Act of 2012 enabled the FAA to take a more stream-lined, risk-based approach to this rule, and to lay a strong foundation that will facilitate safe integration and harness innovation in this rapidly evolving landscape. The flexibility with regard to airworthiness certification for small, low-risk operations that Congress provided in section 333 of the 2012 Act, enabled us to proceed with multiple incremental UAS rules rather than a single omnibus rulemaking.

The public comment period on the proposed small UAS rule is scheduled to close on April 24, 2015. Issuing a small UAS final rule is one of the FAA's and the Department of Transportation's highest priorities, however the timing to promulgate the final rule will depend heavily on the quantity and substance of comments we receive.

BUILDING ON THE FOUNDATION FOR SAFE INTEGRATION OF UAS

The FAA has issued nearly 50 exemptions under section 333 of the 2012 Act and will apply this experience to increase efficiencies and decrease processing time.

The FAA continues to use information and data provided by test sites and other operators, as well as that obtained from its own research and development, or partnerships with other agencies or industry, to continue to identify challenges, validate advanced mitigation strategies, and explore opportunities to progress to the next steps in integrating UAS into the NAS.

Test sites are providing data about the types and sizes of aircraft, number of operations, number of flight hours, notable operating parameters (for example, whether the flight was within or beyond visual line of sight), and any incidents and accidents. Each site has also established its own research agenda. A significant portion of test site data analysis is being performed at the FAA William J. Hughes Technical Center. A Data Lead from the Technical Center, regional representatives, and research engineers, are visiting each UAS test site to evaluate how data is captured and maintained, ensure the integrity of data transferred to the FAA, and determine whether additional data collection would facilitate meeting the FAA's research objectives. The FAA invited public comment in the proposed small UAS rule on how the agency can improve or further leverage its test site program to encourage innovation and safe UAS integration into the NAS.

Consistent with the direction in the agency's FY 2014 appropriation, the FAA is in the process of selecting a new UAS Center of Excellence (COE) which will serve as another resource for identifying solutions for existing and anticipated UAS-related issues. We intend to forge a union of public sector, private sector, and academic institutions to create a world-class consortium that will identify solutions for existing and anticipated UAS-related issues. We are evaluating multiple proposals and plan to establish the COE later this year.

UAS have become increasingly available and affordable to the average consumer, many of whom are not trained aviators. Manned aircraft operators have reported close calls with UAS flying in the airspace. The FAA is taking a proactive approach to educate the public on the safe and responsible use of UAS. The FAA provided model aircraft enthusiasts guidance on the "do's and don'ts" of safe model aircraft operations. Last year, we partnered with members of industry and the modeling community to initiate the "Know Before You Fly" educational outreach campaign that provides UAS operators with the information they need to fly safely and responsibly. The FAA's No Drone Zone initiative, to raise public awareness of the FAA Notice to Airmen, prohibiting unauthorized aircraft- including UAS- from flying over or near NFL regular- and post-season football, games is a success. The No Drone Zone video posted on YouTube prior to the 2015 Super Bowl has received over 57,000 hits, and most importantly, we did not receive any reports of unauthorized activity in the restricted airspace over University of Phoenix Stadium during the game.

Recognizing that local law enforcement is often in the best position to respond quickly, the FAA issued guidance for these first responders to deter, detect, investigate, and report unauthorized or unsafe UAS operations. While our first action is to educate UAS operators about statutory and regulatory compliance, when appropriate, we will use administrative and legal enforcement action to gain compliance.

FUTURE VISION AND CHALLENGES

We are already looking beyond the small UAS rulemaking at what comes next in terms of the types of operations expected, and what technologies we may need to certify. The FAA has consulted with the UAS ARC to determine the next areas on which to focus so as to enable those UAS operations with the highest net societal benefits. These recommendations are being assessed and will result in additional focus areas that will become the centerpiece for FAA's strategic plans for UAS integration.

As the aerospace industry and aviation system grow more complex, we must ensure that our resources are directed to the areas with the highest safety risk. We will need to expand collaborative, data-driven processes with the UAS industry to improve safety and streamline process in areas such as certification. We must meet challenges and take advantage of opportunities.

To reach these objectives, a new advisory circular is being developed to inform the UAS industry how to use a risk based decision-making process to establish certification criteria. This advisory circular is essential for enabling the certification of larger UAS for operation in the NAS.

Another key initiative is one that the FAA is undertaking through a Cooperative Research and Development Agreement with CNN to look at the operations of UAS engaged in newsgathering and at flexible ways to facilitate safe operation over people and in urban areas. These activities will support the development of standards for small UAS intended for use in populated areas. These standards are under development by ASTM International.

The safe integration of UAS into the NAS will be facilitated by new technologies being deployed as part of the Next Generation Air Transportation System (NextGen). NAS Voice System (NVS), Data Communications (Data Comm), and System Wide Information Management (SWIM) will provide more information, flexibility, situational awareness and a greater ability to communicate with NAS users.

The United States has the safest aviation system in the world, and our goal is to integrate this new and important technology while maintaining safety as our highest priority. We are committed to ensuring that the United States continues to lead the world in the development and implementation of aviation technology for safety. We look forward to continuing to work with Congress as we continue to integrate UAS into the NAS.

This concludes my statement. I will be happy to answer your questions at this time.

End Note

[1] RTCA, Inc. is not-for-profit organization that serves as a federal advisory committee to the FAA. See http://www.RTCA.org.

In: Unmanned Aerial Systems
Editor: Wilfred Sparks

ISBN: 978-1-63484-652-3

Chapter 3

TESTIMONY OF JOHN VILLASENOR, NONRESIDENT SENIOR FELLOW, THE BROOKINGS INSTITUTION. HEARING ON "UNMANNED AIRCRAFT SYSTEMS: KEY CONSIDERATIONS REGARDING SAFETY, INNOVATION, ECONOMIC IMPACT, AND PRIVACY"*

Good afternoon Chair Ayotte, Ranking Member Cantwell, and Members of the Subcommittee. Thank you very much for the opportunity to testify today on the important topic of domestic unmanned aircraft systems (UAS).[1]

I am a nonresident senior fellow in Governance Studies and the Center for Technology Innovation at the Brookings Institution. I am also a National Fellow at the Hoover Institution at Stanford, and a professor at UCLA, where I hold appointments in the Electrical Engineering Department and the Department of Public Policy. The views I am expressing here are my own, and do not necessarily represent those of the Brookings Institution, Stanford University or the University of California.

My testimony today can be summarized as follows:[2]

---

* This is an edited, reformatted and augmented version of testimony presented March 24, 2015 before the Senate Committee on Commerce, Science & Transportation, Subcommittee on Aviation Operations, Safety, and Security.

With respect to privacy:

- When considering the possibly of new privacy laws relating to UAS, it is important not to lose sight of the protections we already have. I believe that that our existing legal framework will provide substantially more protection against privacy-violating misuses of UAS than is commonly recognized.
- UAS-specific privacy legislation at the federal or state level must be carefully crafted to avoid unintended consequences. Absent such care, new legislation could inadvertently end up impeding uses of UAS that pose no privacy concerns at all.
- Privacy legislation that would impact non-government UAS users needs to be drafted in a manner that avoids colliding with the First Amendment freedom to gather information.

With respect to the integration of UAS into the National Airspace System:

- Successful integration of UAS will require fundamentally rethinking our approach to managing the airspace below 500 feet above ground level.
- In future years, with proper design and testing, autonomous, beyond-line-of-sight UAS flight can enable important new applications (such as automated search and rescue operations). Congress should provide a mechanism to allow safe testing of these technologies in appropriately selected subareas within the FAA-designated UAS test sites.

With respect to ensuring America's continued technology leadership:

- Robotics will be one of the key technologies of this century. Ensuring that the United States remains a global technology leader will require ensuring that we have a strong robotics industry, and that in turn will require that we maintain leadership in UAS technology. Congress has a central role to play in achieving that goal.
- The unmanned aircraft hobbyist community—which includes both model airplane hobbyists as well as hobbyists who fly what are commonly called "drones"—is a vital pipeline for careers in aviation and technology. It is important to avoid overly narrow regulatory

interpretations that unreasonably limit hobbyists, and that as a consequence also impede America's future capacity for innovation.

UAS: Growing Attention in Congress, and More Broadly

Much has happened in the three years since the FAA Modernization and Reform Act of 2012 (FMRA)[3] was signed into law. As Members of this Subcommittee know well, that law addressed both government (more formally, "public") UAS as well as those operated by non-government (i.e., private and commercial) entities (more formally, "civil"). In particular, under the FMRA, the FAA was directed to "develop and implement operational and certification requirements for the operation of public unmanned aircraft systems in the national airspace system"[4] by the end of 2015. In addition, the FAA was directed to "provide for the safe integration of civil unmanned aircraft systems into the national airspace system as soon as practicable, but not later than" the end of September 2015.[5]

In February 2015, the FAA released a long awaited Notice of Proposed Rulemaking (NPRM)[6] for civil "small" (defined as weighing less than 55 pounds) aircraft. The process of developing these proposed rules had been ongoing since well before the enactment of the FMRA, and their publication marked a significant milestone in developing a civil UAS integration framework.

The past several years have also seen a high level of activity in state legislatures in relation to UAS. According to the National Conference of State Legislatures, UAS-specific laws were enacted in 13 states in 2013[7] and 10 states in 2014.[8]

UAS and Privacy: An Important and Legitimate Concern

It is important to start by acknowledging that the privacy concerns raised by UAS are real and worthy of attention. For the first time ever, UAS are making it easy and inexpensive to obtain overhead imagery. In the coming decade, that capability will be used by dozens of federal government agencies, by hundreds of state and local law enforcement agencies, and by thousands of private companies and individuals.

The overwhelming majority of the time, UAS operators in both the public and private sectors will be mindful of the need to respect privacy. But as the number of UAS users and flights continues to increase, and as imaging technology continues to improve, simple statistics make it inevitable that sometimes, either inadvertently or intentionally, UAS will be used in ways that violate privacy.

That leads to a series of key questions: To what extent are current legal frameworks up to the task of addressing UAS privacy? What new laws, if any, are needed? Should those laws be at the federal or state level, or both? And how can we ensure that any new laws are constitutional and avoid impeding non-privacy-violating uses of UAS?

The February 2015 Presidential Memorandum on UAS Privacy

The UAS privacy question is particularly timely in light of President Obama's February 2015 Presidential Memorandum titled "Promoting Economic Competitiveness While Safeguarding Privacy, Civil Rights, and Civil Liberties in Domestic Use of Unmanned Aircraft Systems" (hereafter, the "Presidential UAS Memorandum").[9] The Presidential UAS Memorandum addresses UAS operated by the federal government and, separately, those operated by commercial and private entities.

With respect to federal government UAS, the Presidential UAS Memorandum provides a series of policies and procedures aimed at protecting privacy and civil liberties and ensuring transparency and accountability. For example, the Presidential UAS Memorandum limits the duration of time that federal agencies can retain information collected using UAS that may contain personally identifiable information.[10] It also requires an agency using UAS to "provide notice to the public regarding where the agency's UAS are authorized to operate in the NAS,"[11] and to "make available to the public, on an annual basis, a general summary of the agency's UAS operations during the previous fiscal year, to include a brief description of types or categories of missions flown, and the number of times the agency provided assistance to other agencies, or to State, local, tribal, or territorial governments."[12]

One important government UAS category not directly[13] addressed by the Presidential UAS Memorandum is state and local government use. The policies identified in the Presidential UAS Memorandum are of necessity limited to federal government agencies using UAS. However, those policies can and should serve as a model for states to consider and potentially adopt, either as is or with modifications.

Separately, with respect to nongovernment (i.e., commercial and private) UAS, the Presidential UAS Memorandum directed the National Telecommunications and Information Administration (NTIA) to initiate a "multi-stakeholder engagement process to develop a framework regarding privacy, accountability, and transparency for commercial and private UAS."[14] This framework will not have the force of law, but will instead be a set of "best practices" that commercial and private UAS operators will presumably be encouraged to adopt. On March 5, 2014, the NTIA published a request for public comment[15] and identified a set of 16 questions relating to privacy, accountability, and transparency in relation to commercial and private UAS. Comments are due to the NTIA on April 20, 2015. The dialog generated during this process will be vital in identifying any loopholes that might exist in current privacy law in relation to non-government UAS, and that could be addressed with suitably crafted legislation.

"PUBLIC NAVIGABLE AIRSPACE"

One of the most interesting and important questions relating to UAS generally, and to the privacy issues they raise specifically, relates to the definition of "public navigable airspace."

Discussions about public navigable airspace in the context of manned aircraft often assert that it is the airspace above 500 feet above ground level. However, that assertion provides an incomplete picture for several reasons. First, it is only partially accurate. Fixed-wing aircraft obviously have a right to use altitudes lower than 500 feet when taking off and landing. In addition, the altitude minimums are higher "[o]ver any congested area of a city, town, or settlement, or over any open air assembly of persons."[16] Furthermore, helicopters are not subject to the same altitude minimums as fixed-wing aircraft as long as "the operation is conducted without hazard to persons or property on the surface."[17]

With UAS the picture gets even more complicated because of rules limiting operation of certain classes of small UAS to a *maximum* of 400 or 500 feet.[18] Clearly, then, when it comes to UAS the public navigable airspace must include some altitudes below 500 feet. But just as clearly, it shouldn't include the airspace two inches above ground in a person's backyard.

In a 1946 case involving manned aircraft (*United States v. Causby*), the Supreme Court recognized the need to provide the public with access to the airspace while also recognizing the need to provide property owners with a zone of control over their land: "We have said that the airspace is a public highway," the Court wrote. "Yet it is obvious that if the landowner is to have full enjoyment of the land, he must have exclusive control of the immediate reaches of the enveloping atmosphere."[19]

But exactly how far up should this exclusive control extend? In a November 2014 paper.[20] Professor Gregory McNeal of the Pepperdine University School of Law provided an excellent framing of the complexities involved. Dr. McNeal observes that one component of a solution could be to give property the owners the right "to exclude aircraft, persons, and other objects from a column of airspace extending from the surface of their land up to 350 feet above ground level."[21] Dr. McNeal also notes that a height limit alone won't be sufficient:

> Granted a rule extending property rights in a manner to prevent low altitude flights directly over a landowner's property won't preclude the police from asking a neighbor if they can fly above their adjacent property to obtain a better vantage point, just like existing rules don't preclude the police (or a private citizen) from asking a neighbor if they can come inside to look out a second floor window into neighboring property. Similarly, such a rule won't preclude the police from flying above public land (such as sidewalks and streets), but local zoning laws could address flights over public land.[22]

To that, I would add the additional concern that codifying the specific boundaries of a property owner's zone of control over airspace would also codify a region (e.g., above 350 feet) in which the property owner does *not* have control. It is easy to envision how this could be exploited. And, with improvements in imaging technology, images acquired from just above the upper limit of a property owner's region of control could still be very invasive.

A more fundamental issue is that while height is certainly one of the factors that impacts whether UAS use over (or in the vicinity of) private property is invasive, it is not the only factor. It is also important to consider what the UAS is doing. A UAS that transits quickly and quietly over a property at 320 feet is generally far less intrusive than one that hovers overhead for many minutes at 380 feet. And a UAS equipped with a very advanced imaging system is potentially much more invasive than one with a very basic imaging system.

Given these complexities, I do not think it is feasible to effectively protect privacy by attempting to codify in advance the specific ways in which it is permissible—or impermissible—to overfly private property. I believe that the better way to address this is to let courts apply tort law and (when applicable[23]) statutory law using the well-established, non-technology-specific standard of a "reasonable expectation of privacy" to the facts specific to any particular case that might arise.

How Protective Is the Current Legal Framework?

Of course, in considering new laws addressing UAS privacy, one of the first questions to ask is: What protections do we *already* have? The answer, I believe, is that our existing legal framework will provide substantially more protection against privacy-violating misuses of UAS than is commonly recognized.

The applicable framework for privacy from UAS depends in large part on who is making the observations. For UAS operated by the government, the Fourth Amendment, which provides the "right of the people to be secure in their persons, houses, papers, and effects, against unreasonable searches and seizures," is a key pillar of privacy protection. For privately operated UAS, privacy protections are provided through the common law invasion of privacy tort as well as through civil and criminal invasion of privacy statutes.

Government UAS and the Fourth Amendment

It is sometimes suggested that because the Fourth Amendment was ratified over 200 years ago, it will not be effective in providing protection from privacy violations using UAS—a technology that the Founders could scarcely have imagined.

I disagree. To explain why, it is helpful to start by considering several *manned* aircraft cases from the 1980s in which the Supreme Court did *not* find a Fourth Amendment violation—and then to consider why, in light of more recent Supreme Court jurisprudence, I believe that the UAS privacy picture is somewhat more optimistic than those precedents might initially appear to suggest.

The Supreme Court's 1986 decision in *California v. Ciraolo*[24] considered police use of a small airplane to overfly a Santa Clara, California residence at 1000 feet and look into the backyard, where they saw marijuana plants. When presented with the question of whether the observations violated the Fourth Amendment, the Supreme Court found in favor of the government, writing that because the observations of the curtilage of the respondent's home were made from "public navigable airspace ... in a physically nonintrusive manner," the respondent's expectation of privacy from such aerial observations was not one "that society is prepared to honor."[25] The Court concluded that "[i]n an age where private and commercial flight in the public airways is routine, it is unreasonable for respondent to expect that his marijuana plants were constitutionally protected from being observed with the naked eye from an altitude of 1,000 feet."[26]

Also in 1986, in a ruling involving government overflights of an industrial facility, the Court found in *Dow Chemical Co. v. United States* that "the taking of aerial photographs of an industrial plant complex from navigable airspace is not a search prohibited by the Fourth Amendment."[27] The Court in *Dow Chemical* considered the open areas in the 2000-acre industrial facility more akin to an "open field" than to the curtilage of a home, and concluded that those areas were "open to the view and observation of persons in aircraft lawfully in the public airspace immediately above or sufficiently near the area for the reach of cameras."[28] And in 1989, in *Florida v. Riley*,[29] a case similar in some respects to *Ciraolo*, the Supreme Court again considered the constitutionality of aerial observations of a home's curtilage by law enforcement. A majority of the justices in *Riley* found the observations constitutional.[30]

In combination, these rulings certainly suggest that some observations from government UAS will be deemed constitutional. However, and critically, that does not mean that *all* such observations will be constitutional. If the government uses a UAS, without a warrant, in a manner violating a reasonable expectation of privacy—either through the duration of the observations or the detail they reveal—then those observations should not pass constitutional muster.

We don't yet know how the Supreme Court would rule in a case involving UAS privacy, but a review of the Court's 21st century jurisprudence in relation to other technologies provides cause for optimism. In 2001, for example, the Supreme Court ruled in *Kyllo v. United States*[31] that the warrantless use by the police of a thermal imaging camera to measure the temperature of the walls of a house—and to thereby infer that the occupant was growing marijuana— was a violation of the Fourth Amendment.

In 2014, the Court ruled in *Riley v. California*[32] that when conducting a search incident to arrest, police did not generally have the right to search the contents of the arrestee's mobile phone without a warrant. Writing for the Court, Chief Justice Roberts explained:

> Our cases have recognized that the Fourth Amendment was the founding generation's response to the reviled "general warrants" and "writs of assistance" of the colonial era, which allowed British officers to rummage through homes in an unrestrained search for evidence of criminal activity.[33]

And:

> Modern cell phones are not just another technological convenience. With all they contain and all they may reveal, they hold for many Americans "the privacies of life". The fact that technology now allows an individual to carry such information in his hand does not make the information any less worthy of the protection for which the Founders fought. Our answer to the question of what police must do before searching a cell phone seized incident to an arrest is accordingly simple— get a warrant.[34]

The 2012 *United States v. Jones*[35] decision also sheds light on how some of the Justices view the Fourth Amendment in light of modern technologies. That case considered the government's installation, without a valid warrant, of a GPS tracking device on a vehicle used by a suspect in a narcotics investigation. The Court's decision was unanimous in finding the government's actions unconstitutional, but there was considerable divergence in the basis for that finding. The majority opinion, delivered by Justice Scalia, found a Fourth Amendment violation in the physical trespass that occurred during the placement of the GPS device on the vehicle. That intrusion, wrote Justice Scalia, "would have been considered a 'search' within the meaning of the Fourth Amendment when it was adopted."[36]

In terms of viewing the Fourth Amendment in light of today's technologies, the concurrences in *Jones* were more instructive than the majority opinion. In a concurrence joined by three other Justices, Justice Alito wrote that the question is whether the "respondent's reasonable expectations of privacy were violated by the long-term monitoring"[37] of his vehicle. Because "law enforcement agents tracked every movement that respondent made in the vehicle he was driving"[38] for four weeks—a level of monitoring that Justice Alito felt impinged on reasonable expectations of privacy—Justice Alito concluded that the tracking constituted a search.

Justice Sotomayor, in addition to joining the majority, provided a separate concurring opinion arguing that "the trespassory test ... reflects an irreducible constitutional minimum"[39] and agreeing with Justice Alito's view that the respondent's reasonable expectations of privacy were violated. Justice Sotomayor also expressed concern that the unchecked ability of the government to assemble "the sum of one's public movements" could enable it to obtain private information regarding political and religious beliefs.[40]

So we have a total of five Justices—Justice Alito and the three others[41] who joined his concurrence, and Justice Sotomayor in her own concurrence—on record with statements indicating a view that warrantless use of technology to perform long-term tracking violates the Fourth Amendment. This is relevant to UAS both directly and indirectly.

The direct relevance arises because there is one class of UAS, called "High Altitude, Long Endurance" (HALE) UAS, that can stay aloft at very high altitudes for weeks, months, or even years at a time. The concurrences in *Jones* suggest that government use of such platforms to perform warrantless long-term tracking of individuals using this or any other technology would raise serious constitutional concerns for a majority of the Justices.

There is also an indirect and far broader relevance that is not specific to UAS. After all, most UAS can only stay aloft only for short periods of time—usually measured in minutes, not hours. They simply cannot be used to perform long-term tracking. They can potentially, however, be misused in ways that would violate reasonable expectations of privacy. The concurrences in *Jones*—as well as the majority opinions in *Kyllo* and in *Riley v. California*—indicate that the Fourth Amendment, when properly interpreted, retains the power to prevent the government from using modern technologies in ways that violate privacy.

Non-Government Unmanned Aircraft and Privacy

Private entities are not bound by Fourth Amendment restrictions that apply to the government and have an affirmative First Amendment privilege to gather information. However, while that privilege is extensive, it ends when it crosses into an invasion of privacy.

Use of a UAS to invade an individual's privacy could result in civil or criminal liability. With respect to civil liability, courts in most jurisdictions recognize the two forms of common law invasion of privacy most likely to arise in connection with UAS: intrusion upon seclusion and public disclosure of private facts.[42] In addition, some states also have civil or criminal statutes, or both, related to invasion of privacy.[43]

State laws aimed at bolstering privacy protections from non-government UAS should be enacted only if, and to the extent that, current frameworks are insufficiently protective. After all, the set of existing protections against invasion of privacy can be powerful and adaptable precisely because they are not technology-specific, and can therefore be reinterpreted as needed as new technologies emerge. Technology-specific privacy laws, by contrast, risk becoming quickly obsolete as the technology changes. Alternatively, or in addition, they can sometimes lead to unintended consequences that impede uses of the technology that pose no threat at all to privacy.

THE IMPORTANCE OF FIRST AMENDMENT CONSISTENCY

There is also an additional consideration to keep in mind: Laws drafted to address privacy with respect to a specific, rapidly evolving technology such as UAS can inadvertently run counter to the goal of a technology-neutral interpretation of the First Amendment freedom to gather information. This can create some problematic consequences.

Consider a photograph of the countryside taken using a smartphone by a passenger[44] riding in a privately owned single-engine airplane as it descends through 350 feet on the way to landing. No one would reasonably deny the passenger's First Amendment right to take that photograph. The owners of properties within the frame of view would not have an ownership interest in the photograph; nor would they have the right to control its use, to require notification that the photograph had been acquired, or to require that it be retained for a certain amount of time.

Now consider a similar photograph taken from a privately-owned UAS 350 feet above the ground, using a camera with similar imaging capabilities.[45] Under some of the UAS privacy laws that have been proposed (and in some states, introduced and adopted), the owners of properties within the frame of view might have substantially more control over the acquisition, use, dissemination, or retention of the UAS-acquired photograph than of the photograph taken by the on-board passenger.

I will emphasize that neither the UAS operator nor the on-board passenger has the right to acquire images that constitute an invasion of privacy. But, assuming that the images are such that no one's privacy is invaded (for example, if the image resolution is low and no private details are contained in the images), why should the UAS operator's right to take a photograph be so much more limited than that of the on-board passenger? Put another way, why should the scope of the First Amendment be so much narrower for the UAS operator than for the on-board passenger?

If the First Amendment is in effect narrowed for a particular class of people—UAS users, in this instance—it is easy to see how this could lead to some disturbing unintended consequences. Peaceful demonstrators, for example, might be told that they are not permitted to use a UAS to film a demonstration, on the grounds that the footage might include adjacent buildings owned by people who disagree with their viewpoint.

INTEGRATING UAS INTO THE NAS: MEETING THE CHALLENGES

Without in any way diminishing the importance of the issue of UAS privacy, I would also like to offer some more general comments regarding the integration of UAS into the NAS.

The airspace in the United States is a complex, busy place. It is shared every day by thousands of manned aircraft, including single-engine private planes flying at 100 miles per hour and 500-ton commercial passenger aircraft[46] travelling at well over 500 miles per hour. At this moment, and in fact at most times during most days, there are many thousands of manned aircraft in the air over the United States.[47] We take it for granted that nearly all of the time, all of these aircraft share the airspace without incident, in all types of weather, day and night. Anyone who spends a few minutes watching the radar tracks of airplanes above a major U.S. metropolitan area will come away amazed by the complex, three-dimensional choreography involved in keeping our skies safe.

To this already complicated mix, we will be adding thousands of new unmanned aircraft. For UAS integration to occur as safely and successfully as possible, we will need to rethink management of the airspace below 500 feet above ground level. In addition, we should recognize that autonomous flight can play an important longer-term role. And, we should provide a mechanism to enable safe testing of autonomous, beyond-line-of-sight UAS operation.

THE AIRSPACE BELOW 500 FEET: THE NEED FOR A NEW APPROACH

We need to rethink management of the airspace below 500 feet above ground level. The paradigms developed for manned flight that generally occurs above 500 feet are much less well suited to unmanned flight that will occur below 500 feet. Instead, we need a new approach that recognizes 1) the important stake of property owners in how sub-500-foot altitudes will be managed and used, 2) the fact that UAS will outnumber manned aircraft at those altitudes, and 3) that since so much of the low-altitude traffic will be unmanned, the traditional assumption that every aircraft must have at least one human pilot devoting his or her full attention to flying it will, in the long term, need to be revisited. This last point ties directly to the role of autonomous flight technologies that, with proper design, have the potential to improve both safety and efficiency in the lower airspace.

AUTONOMOUS FLIGHT

"Autonomous" UAS flight refers to a UAS that is flown without being actively and continuously controlled by a ground-based human pilot.[48] Autonomy is actually a continuum, including fully autonomous flight as well as flight that is mostly under the control of a human pilot. It is also possible to have a UAS that flies autonomously during some, but not all, portions of a flight. (In some respects, this isn't so different from what routinely occurs today in the context of manned flight using technologies like autopilot.) In discussions about UAS integration, there is often an assumption that autonomous unmanned UAS will pose more dangers than human-piloted UAS. I believe that is an oversimplification.

Certainly, autonomous UAS flight without the proper safeguards would pose very real safety risks. But when the algorithms used to control flight are designed with sufficient care and properly tested, autonomous flight has the potential to deliver very important benefits, including enabling new applications such as the automated search and rescue scenario that I discuss below.

AUTONOMOUS, BEYOND-LINE-OF-SIGHT UAS OPERATION

UAS flight that is both autonomous and beyond-line-of-sight (BLOS) is another important area of technology development. BLOS refers to operation in which the UAS cannot be seen by a person overseeing its operation—either because it is obscured by intervening objects such as trees or buildings, or because it is over the horizon.[49]

BLOS and autonomous flight are not necessarily coupled. Today's technology makes it possible (though it is not generally permitted[50] in the current regulatory environment) for a ground-based human pilot to fly a UAS beyond the line of sight using a computer, console, or other display system showing live video from a UAS-mounted camera. The pilot sees what he or she would see from onboard the UAS, and can navigate the aircraft accordingly, despite not being able to actually see the UAS from his or her location on the ground.[51] This is an example of BLOS flight that is *not* autonomous. Conversely, autonomous UAS flight could be performed in very close proximity to—and within view of—the person who initiated the flight. This is an example of autonomous flight that is *not* BLOS.

But some of the most compelling future applications of UAS involve flight that is *both* autonomous *and* BLOS. Consider the challenge of looking for a lost hiker in the wilderness at night, several miles from the nearest road. It could be very helpful to be able to program a UAS to fly autonomously to the search area, and then to fly automated search patterns, using a thermal imager to identify heat signatures that may indicate the lost hiker. Upon finding a possible match, the UAS could then alert a human pilot, who would then retake control and use the UAS to perform a closer inspection to see if the hiker had indeed been located. To take this example one step further, it would be possible to have half a dozen UAS collaboratively perform a search. This would allow searches to be conducted much faster and much more cost effectively than is possible today using manned aircraft.

Under current regulations (as well as those proposed in the recent NPRM), a UAS flight operation of this sort would not be permissible.[52] In addition, it is currently very difficult for companies or government agencies interested in developing this capability to even *test* these capabilities. Rules at the six UAS test sites recently chosen by the FAA currently prohibit flight that is simultaneously BLOS and autonomous. This forces developers of this technology to either limit their testing to the small confines of indoor spaces, or to test overseas in a country where the rules relating to autonomous UAS flight offer more flexibility.

Today, UAS technology is not yet sufficiently mature to allow autonomous, beyond-line-of-sight UAS operations in the regular (outside of suitable test sites) airspace. But I think it is important to move towards a regulatory framework that could provide a mechanism to safely test and refine these technologies. This could occur, for example, in regions within the already-designated FAA test sites and/or on large parcels of private property owned or leased by the company performing the tests.

Promoting Innovation and Economic Prosperity

In the coming years, UAS will be used to improve agricultural yields, to perform scientific research, to spot and fight forest fires, to perform search and rescue, and to support disaster response. Developing the UAS technologies to enable these and many other applications will involve new business models, new companies, and new jobs. But that will only occur if there is a regulatory climate that fosters a thriving community of companies and individuals with an interest in investing their time, money, and energy to create the safest, most innovative unmanned aircraft technologies of the future.

While there has been plenty of attention to the important issue of commercial UAS, I would also like to highlight the importance of the unmanned aircraft hobbyist community. This community, which includes both model airplane hobbyists as well as those who fly what are commonly called “drones,” provides a critically important pipeline for careers in aviation and technology. I would urge the FAA and Congress to ensure that as we navigate the complex process of making and interpreting rules for unmanned aircraft, we keep in mind the vital importance of the hobbyist community for our future innovation capacity.

Thank you again for the opportunity to testify on this important topic.

End Notes

[1] The acronym "UAS" is also sometimes expanded to "unmanned aerial systems."

[2] A small portion of this written testimony is excerpted from John Villasenor, Observations From Above: Unmanned Aircraft Systems and Privacy, 36 Harv. J.L. & Pub. Pol'y 457 (2013) and from Eyes in the Sky: The Domestic Use of Unmanned Aerial Systems, Written Testimony of John Villasenor before the House Committee on the Judiciary - Subcommittee on Crime, Terrorism, Homeland Security, and Investigations (May 13, 2013), available at http://www.brookings.edu/~/media/research/files/testimony/2013/05/17%20privacy%20drones%20villasenor/villasenortestimonymay17.

[3] Pub. L. No. 112-95, § 331, 126 Stat. 11, 72 (2012)

[4] Id. § 334(b), 126 Stat. at 76.

[5] Id. § 332(a)(3).

[6] Operation and Certification of Small Unmanned Aircraft Systems, 80 Fed. Reg. 9544 (Feb. 23, 2015), available at http://www.faa.gov/ regulations_policies/rulemaking/recently_ published/media/2120-AJ60_NPRM_2-15-2015_joint_signature.pdf.

[7] 2013 Unmanned Aircraft System (UAS) Legislation, NCSL.Org, http://www.ncsl.org/research/civil-and-criminal-justice/unmanned-aerialvehicles.aspx (last visited on Mar. 15, 2015).

[8] Rich Williams, 2014 State Unmanned Aircraft System (UAS) Legislation, NCSL.ORG (Sept. 16, 2014), http://www.ncsl.org/research/civiland-criminal-justice/2014-state-unmanned-aircraft-systems-uas-legislation.aspx.

[9] The White House, Office of the Press Secretary, Presidential Memorandum: Promoting Economic Competitiveness While Safeguarding Privacy, Civil Rights, and Civil Liberties in Domestic Use of Unmanned Aircraft SystemsWHiTE HOuSE.GOv (Feb. 15, 2015), available at https://www.whitehouse.gov/the-press-office/2015/02/15/presidential-memorandum-promoting-economic-competitiveness-while-safegua.

[10] Id. at §1(a)(ii).

[11] Id. at §1(d)(i).

[12] Id. at §1(d)(iii).

[13] The Presidential UAS Memorandum indirectly addresses state and local government use by requiring that "State, local, tribal, and territorial government recipients of Federal grant funding for the purchase or use of UAS for their own operations have in place policies and procedures to safeguard individuals' privacy, civil rights, and civil liberties prior to expending such funds." See the Presidential UAS Memorandum at §1(c)(vi).

[14] Id. at §2(b).

[15] Privacy, Transparency, and Accountability Regarding Commercial and Private Use of Unmanned Aircraft Systems, 80 Fed. Reg. 11978 (Mar. 5, 2015), available at http://www.ntia.doc.gov/files/ntia/publications/rfc_uas_privacy_03052015.pdf.

[16] 14 C.F.R. §91.119(b).

[17] 14 C.F.R. §91.119(d).

[18] The NPRM released in February 2015 (supra note 8) identifies a proposed upper limit of 500 feet for small (up to 55 pounds) UAS and an upper limit of 400 feet for the "Micro UAS" sub-classification, which covers UAS up to 4.4 pounds.

[19] United States v. Causby, 328 U.S. 256, 264 (1946).

[20] Gregory McNeal, Drones and Aerial Surveillance: Considerations for Legislators, The Brookings Institution (Nov. 2014), available at http://www.brookings.edu/research/reports2/2014/11/drones-and-aerial-surveillance.

[21] Id. at 4.

[22] Id. at 13.

[23] Some invasion of privacy statutes codify a "reasonable expectation of privacy" as the standard to use when judgeing whether the statute has been violated.

[24] 476 U.S. 207 (1986).

[25] Id. at 213–14.

[26] Id. at 215.

[27] Dow Chem. Co. v. United States, 476 U.S. 227, 239 (1986).

[28] Id.

[29] 488 U.S. 445 (1989).

[30] The 1989 Riley decision comprised an opinion delivered by Justice White and joined by three other Justices and an opinion from Justice O'Connor concurring in the judgment. Thus, while there was no majority opinion, a majority of the Justices found the observations constitutional.

[31] 533 U.S. 27 (2001).

[32] 134 S. Ct. 2473 (2014).

[33] Id. at 2494.

[34] Id. at 2494-2495, internal citations omitted.

[35] 132 S. Ct. 945 (2012).

[36] Id. at 949.

[37] Id. at 958 (Alito, J., concurring).

[38] Id. at 964.

[39] Id. at 955 (Sotomayor, J., concurring).

[40] Id. at 956.

[41] Justice Alito's concurrence was joined by Justices Ginsburg, Breyer, and Kagan.

[42] See William L. Prosser, Privacy, 48 Calif. L. Rev. 383, 389 (1960); see also Restatement (Second) Of Torts §§ 652A–652E (1997).

[43] See, e.g., CAL. Civ. CODE § 1708.8 (West 2011).

[44] I am assuming in this example that the passenger is not a government employee or otherwise acting on behalf of a government entity.

[45] Of course, a UAS would not be able to operate in the immediate vicinity of an airport, which is why the photograph could be "similar" but could not be "identical."

[46] The gross takeoff weight of an Airbus A380 is over one million pounds.

[47] Air Traffic NextGen Briefing: Keeping America's Skies Safe, FAA.Org (last modified Sept. 18, 2014), https://www.faa.gov/air traffic/ briefing/ (last visited on Mar. 15, 2015).

[48] Autonomy is actually a continuum, depending upon the level of control over the flight maintained by an on-the-ground human pilot.

[49] I am referring here to beyond the visual line of sight, in which the UAS operator cannot see the UAS. In some other contexts, the term "line of sight" is used to in association with whether or not there is direct radio communication with the UAS, without any need to relay the radio signal through an intermediate location.

[50] There is an exception under which, in regions of arctic Alaska, certain beyond-line-of-sight UAS operations are permitted. See http:// www.faa.gov/news/updates/?newsId=73981

[51] This type of flight is called "first person view" or "FPV."

[52] Under the NPRM, operators of "small" UAS would be required to be "visually capable of seeing the small UAS", i.e., BLOS operations would not be permitted.

In: Unmanned Aerial Systems
Editor: Wilfred Sparks
ISBN: 978-1-63484-652-3

Chapter 4

TESTIMONY OF PAUL MISENER, VICE PRESIDENT OF GLOBAL PUBLIC POLICY, AMAZON, INC. HEARING ON "UNMANNED AIRCRAFT SYSTEMS: KEY CONSIDERATIONS REGARDING SAFETY, INNOVATION, ECONOMIC IMPACT, AND PRIVACY"*

Thank you, Chairwoman Ayotte and Ranking Member Cantwell. My name is Paul Misener, and I am Amazon's Vice President for Global Public Policy. Unmanned aircraft systems ("UAS") present tremendous opportunities for innovation and soon will provide consumer services unimagined only a decade ago. Thank you for your attention to this important topic; for calling this hearing; and for inviting me to testify.

I. AMAZON PRIME AIR

Amazon Prime Air is a future service that will deliver packages to customers in 30 minutes or less using small UAS. Flying below 500 feet, and

* This is an edited, reformatted and augmented version of testimony presented March 24, 2015 before the Senate Committee on Commerce, Science & Transportation, Subcommittee on Aviation Operations, Safety, and Security.

generally above 200 feet except for takeoff and landing, and weighing less than 55 pounds total, Prime Air UAS will take advantage of sophisticated "sense and avoid" technology, as well as a high degree of automation, to ensure safe operations including at distances of 10 miles or more, well beyond visual line of sight.

Not only do we think our customers will love this service, we believe it will benefit society more broadly. Once operational, Prime Air will increase the overall safety and efficiency of the current ground transportation system, by allowing people to skip the quick trip to the store or by decreasing package delivery by truck or car. For the same reasons, Prime Air will reduce buyers' environmental footprint: if a consumer wants a small item quickly, instead of driving to go shopping or causing delivery automobiles to come to her home or office, a small, electrically-powered UAS will make the trip faster and more efficiently and cleanly. To realize these tremendous opportunities for innovation and other benefits in the United States, additional research and development – and, soon, rules of operation – are needed.

Amazon Prime Air has been conducting outdoor R&D flight testing in multiple locations abroad, *i.e.*, in other countries. Our testing abroad has required but minimal aviation regulatory approval, given the low risk presented by our small UAS designs; the R&D nature of our flight activity; and our relatively rural test sites. Nowhere outside of the United States have we been required to wait more than one or two months to begin testing, and permission has been granted for operating a *category* of UAS, giving us room to experiment and rapidly perfect designs without being required to continually obtain new approvals for specific UAS vehicles. Our outdoor flight testing is going well, and we are very pleased with the R&D progress this testing has enabled.

In addition to this work, we also will prepare our distribution network for the eventual integration of Prime Air delivery service. Preparation will include optimizing our internal systems because, in order to meet our Prime Air customer delivery goal of 30 minutes or less, our UAS must be loaded quickly, and this presents fascinating logistical challenges, including within our huge warehouses.

II. International Regulatory Activities

No country in which we now have distribution facilities has yet adopted rules that would allow commercial UAS package deliveries. So, in addition to our Prime Air R&D; testing; and distribution network preparations, we are working with government agencies to develop appropriate rules for small UAS operations. Such rules must allow UAS applications to take advantage of a core capability of the technology: to fly with minimal human involvement, beyond visual line of sight. Such rules of operation should be proportionate to risk, setting a level of safety but not mandating how that level must be met.

UAS present obvious safety risks of mid-air collisions and crashes to the ground. In its recently-released Notice of Proposed Rulemaking ("NPRM"), the US Federal Aviation Administration ("FAA") tied these concerns to the lack of human "see and avoid" abilities and the hazard of ground-to-air communications "link loss." Both of these factors would have been difficult to address even just a decade ago, but automated UAS *sense* and avoid technology and on-board intelligence address these factors and will mitigate the related risks.

Safety is Amazon's top priority and, earlier this month, I discussed UAS safety with Europe's most senior leaders of aviation regulation, including at the UK's Department for Transport and Civil Aviation Authority; the European Commission ("EC"); and the European Aviation Safety Agency ("EASA"). I also conferred with leadership of the Joint Authorities for Rulemaking on Unmanned Systems ("JARUS"), in which Asian, European, and North American aviation officials are working on a set of UAS technical, safety, and operational requirements to be recommended to aviation authorities worldwide. Amazon also participated in the EC conference on UAS in Riga, Latvia, and this week we are attending the International Civil Aviation Organization ("ICAO") UAS conference in Montreal, Canada.

I'm delighted to report that these aviation authorities with whom we met in the UK and at the multinational bodies are enthusiastically pursuing regulatory frameworks and operational rules for UAS. The approach they are taking is eminently reasonable: it is risk- and performance-based, and it is mindful of the tremendous opportunities for innovation and economic benefits that UAS present.

Two reports released this month, from the EC's Riga conference and from EASA, demonstrate the important planning already underway for future commercial UAS operations. The official report of the Riga conference included several important conclusions: UAS should be treated as new types of aircraft with proportionate rules based on the risk of the operation ("rules should be simple and performance based"); rules must be developed now ("the basic regulatory framework should be put in place without delay"); technologies and standards need to be developed for the full integration of

UAS in the airspace; and EASA should lead the harmonization of UAS regulation across Europe. Following the EC's Riga conference, EASA presented its new regulatory approach for UAS operations, concluding that:

> The operation of [UAS] should be regulated in a manner proportionate to the risk of the specific operation. Considering the broad range of operations and types of [UAS], it is proposed to establish 3 categories of operations and their associated regulatory regime.... This concept has been developed to address two main goals: (a) Integration and acceptance of [UAS] into the existing aviation system in a safe and proportionate manner; [and] (b) Foster an innovative and competitive European [UAS] industry, creating new employment, in particular for SMEs.

By the end of 2015, EASA plans to present a "draft regulatory framework" to the European Commission.

III. Testing and Planning in the United States

A. Current Rules for R&D Testing

American commercial entities want to innovate and perfect UAS technology, and to do so we must conduct R&D testing. Amazon has a large *indoor* R&D facility in Seattle. In this facility, our Prime Air team of roboticists, scientists, aeronautical engineers, remote sensing experts, and a former NASA astronaut has conducted flight tests on rapidly improving designs. But of course we need to safely test these designs *outdoors*, exposed to the flight conditions our UAS eventually will experience in operations – namely, wind, turbulence, and the variety of temperature, humidity, and precipitation conditions of the real world.

So, beginning in early 2014, we began talking to the FAA about obtaining permission to conduct R&D testing outdoors. And, from the beginning, we made clear that the rapid pace of UAS innovation means that we need permission to rapidly modify our test vehicles, without administrative delays associated with every change.

We are very grateful to the FAA for granting us permission to conduct UAS testing outdoors in the United States. This approval came last Thursday, and we're eager to get flying here as we have been abroad. However, the permission the FAA granted is more restrictive than are the rules and approvals by which we conduct outdoor testing in the UK and elsewhere. (It's even more limited than the rules applicable to non-commercial, amateur UAS fliers in the United States.) Moreover, obtaining permission took far too long, and certainly much longer – over half a year – than it took in other countries.

The good news is that, while the FAA was considering our applications for testing, we innovated so rapidly that the UAS approved last week by the FAA has become obsolete. We don't test it anymore. We've moved on to more advanced designs that we already are testing abroad. Last Friday, we asked the FAA for permission to fly one of these advanced UAS in the United States, as well, and we are hopeful that this permission will be granted quickly.

B. Planning for Future Operations

Although the United States is catching up in permitting current commercial UAS testing, the United States remains behind in planning for future commercial UAS operations.

We are grateful for the FAA's newly-released NPRM, so far as it goes. But it doesn't go far enough. Unlike the planning by the national and multinational groups with whom I met in Europe earlier this month, the FAA is not adequately addressing compelling UAS applications that involve highly automated operations beyond visual line of sight. The FAA has proposed rules in the NPRM – to be adopted probably in 18-24 months – that simply do not address these extremely important applications. The NPRM only briefly requests comments on *whether* the rules should permit operations beyond visual line of sight and, if so, how enabling technology should be evaluated.

Although the FAA has asked a subcommittee of one of its industry advisory committees to examine beyond visual line of sight operations (and I am a member of this subcommittee), the group has only met twice since its inception last year. This low level of government attention and slow pace are inadequate, especially compared to the regulatory efforts in other countries. This is not to suggest that regulators here or abroad can quickly adopt actual rules for UAS operations beyond visual line of sight. That may take some time. But surely regulators should start *proposing* regulatory frameworks and rules for future commercial UAS operations now.

IV. Opportunities for FAA and Congressional Action

Because the United States remains behind in planning for future commercial UAS operations, one might assume that Congress must step in to provide the FAA authority to act. But the fact is that, with few exceptions, the agency already has adequate statutory authority. What the FAA needs is *impetus*, lest the United States fall further behind.

Any impetus embraced by, or given to, the FAA should result in the agency commencing – now – to plan and develop rules for UAS operations that would encompass highly automated flights, beyond visual line of sight. A good starting point could be the ongoing work in the multinational body JARUS, in which an FAA staffer serves as the vice chair. Elevating the level and intensity of FAA participation in this group is one way the United States could confirm its commitment to UAS technology and services. And, here at home, the FAA could immediately begin – or be directed to begin – proposing regulatory frameworks and rules for operations of highly automated UAS beyond visual line of sight, perhaps through a further notice of proposed rulemaking.

Consumer privacy is an area in which the US approach to UAS regulation already is particularly strong. We recognize that UAS technology could cause privacy infringement if commercial operations are not undertaken in a sensible, privacy-conscious manner. Prime Air is a future delivery service, not a surveillance operation, and we will respect the privacy of every person, with stringent privacy policies accessible to all. We strongly support the Commerce Department's effort to develop, through a multi-stakeholder process, best practices on privacy, transparency, and accountability.

Lastly, international harmonization of rules is strongly desirable, and domestic balkanization by states and localities is not. Harmonized rules, perhaps developed through JARUS or ICAO, should be a top FAA priority internationally. And within the United States, uniform federal rules should apply.

V. Conclusion

In conclusion, Madam Chair, I look forward to working with you, your Subcommittee, and the FAA to ensure that important commercial UAS services become available in the United States safely and soon. And I welcome your questions.

In: Unmanned Aerial Systems
Editor: Wilfred Sparks
ISBN: 978-1-63484-652-3

Chapter 5

STATEMENT OF JEFF VANDERWERFF, VWO, LLC., REPRESENTING THE AMERICAN FARM BUREAU FEDERATION. HEARING ON "UNMANNED AIRCRAFT SYSTEMS: KEY CONSIDERATIONS REGARDING SAFETY, INNOVATION, ECONOMIC IMPACT, AND PRIVACY"*

The American Farm Bureau Federation (Farm Bureau) is the nation's largest general farm organization, representing agricultural producers of nearly every type of crop and livestock across all 50 states and Puerto Rico. We appreciate the opportunity to submit a statement to the Subcommittee on Aviation Operations, Safety, and Security for this hearing on Unmanned Aircraft Systems: Key Considerations Regarding Safety, Innovation, Economic Impact, and Privacy.

Farm Bureau supports the use of unmanned aircraft systems (UAS) as another tool for farmers and ranchers to use in managing their crops and livestock and making important business decisions. A farmer faces daily challenges that can affect the farmer's yield, environmental conditions on the farmer's property and, ultimately, the economic viability of the farm. Farmers rely on accurate data to make these decisions and the use of UAS adds a

* This is an edited, reformatted and augmented version of a statement presented March 24, 2015 before the Senate Committee on Commerce, Science & Transportation, Subcommittee on Aviation Operations, Safety, and Security.

valuable and accurate tool for the farmer in making optimal decisions to maximize return on farming operations.

It has become widely accepted that the introduction of UAS into the commercial industry will begin with American agriculture. The primary reason American agriculture is viewed as the pioneer industry to use UAS for a commercial purpose is that the airspace above the fields used in agriculture is low risk, and many of the fields are located in remote areas.

Farm Bureau sees another reason American agriculture will pioneer this effort. America's farmers and ranchers embrace technology that allows their farming businesses to be more efficient, economical and environmentally friendly. American agriculture continues to evolve, and today's farmers and ranchers are using precision agricultural techniques to make business decisions. These decisions can impact the amount of fertilizer a farmer needs to purchase and apply to the field; the amount of water needed to sustain the crop; and the amount and type of herbicides or pesticides the farmer may need to apply. These are only a few examples of the business decisions a farmer makes on a daily basis to achieve optimal yield, lower environmental impact and maximize profits.

Farmers and ranchers are excited to see the transformation of a hobbyist activity into the newest tool for precision agriculture. The U.S. Department of Agriculture defines precision agriculture as "a management system that is information and technology based, is site specific and uses one or more of the following sources of data: soils, crops, nutrients, pests, moisture, or yield, for optimum profitability, sustainability, and protection of the environment."[1] This definition encompasses the purpose of UAS within the agriculture industry.

Farm Bureau sees the benefit of UAS through their ability to provide detailed scouting information on weed emergence, insect infestations and potential nutrient shortages. This valuable information allows the farmer to catch these threats before they develop into significant and catastrophic problems. By addressing threats quickly, the farmer has a greater likelihood of being able to respond appropriately so as to optimize yields.

The imagery from UAS also allows the farmer to spot-treat sections of the fields as opposed to watering or spraying the entire field. The quicker a farmer can discover a potential threat, the quicker the farmer can address the issue. Images from UAS allow the farmer to identify the specific location where a specific treatment – be it fertilizer, water, pesticides or herbicides – is necessary; in doing so, the farmer can eliminate the need to use these applications more broadly across the entire field. By spot-treating threats to the crop, the farmer not only lowers the cost of treatment but also has the potential of lowering the environmental impact by minimizing application.

Farm Bureau is glad to see the Federal Aviation Administration (FAA) introduce its notice of proposed rulemaking for small-UAS. Farm Bureau is in the process of developing its comments to the FAA regarding the proposed rule. It is our hope that farmers and ranchers are able to secure the rights through this process to use UAS as part of their precision agricultural systems. That will allow them to scout fields and will serve as another tactic at their disposal to limit the use of agricultural inputs to only those areas of the field that require treatment. That will be good for the environment as we will be able to grow more with less. Many farmers will adopt this technology as yet another way to live up to the promise of continuous improvement in food production.

While Farm Bureau supports this new technology and the potential opportunities it offers for farmers and ranchers, Farm Bureau is also concerned about the data collected from UAS and the privacy and security of that data.

Even if an individual operator follows all the applicable rules, regulations, and best management practices in his or her farming operation, there is still concern that regulatory agencies or one of the numerous environmental organizations that unnecessarily targets agriculture might gain access to individual farm data through subpoenas. While a farmer's pesticide or biotech seed usage may be a necessary and accepted practice, it could also be politically unpopular with certain groups.

The biggest fear that farmers face in data collection is government accessing their data and using it against them.

Questions abound within the agricultural community about "who owns and controls the data." If a farmer contracts with a company authorized to fly UAS, does the farmer own all the data from that UAS or is it shared by both the contractor and the farmer? In the case of a farm on rented ground, does the tenant or the landlord own the data?

Farm Bureau supports the use of UAS and believes it will be an important addition to farmers' management toolbox, but it is critical that the data remain under the ownership and control of the farmer and is not available to government agencies or others without permission.

End Note

[1] USDA, Natural Resources Conservation Service, "Precision Agriculture: NRCS Support for Emerging Technologies." http://www.nrcs.usda.gov/Internet/FSE DOCUMENTS/stelprdb 1043474.pdf. June 2007.

In: Unmanned Aerial Systems
Editor: Wilfred Sparks

ISBN: 978-1-63484-652-3

Chapter 6

SUMMARY OF SUBJECT MATTER FOR THE HEARING ON "ENSURING AVIATION SAFETY IN THE ERA OF UNMANNED AIRCRAFT SYSTEMS"*

Subcommittee on Aviation Staff

PURPOSE

The Subcommittee on Aviation will meet on Wednesday, October 7, 2015 at 10:00 a.m. in 2167 Rayburn House Office Building to explore issues related to aviation safety as the number of unmanned aircraft systems (UAS) increases in the United States. The Subcommittee will receive testimony from the Federal Aviation Administration (FAA), the United States Forest Service, the Academy of Model Aeronautics (AMA), the Air Line Pilots Association (ALPA) and a Professor of Aeronautics and Astronautics.

* This is an edited, reformatted and augmented version of a preparatory document issued October 2, 2015 by the House Committee on Transportation and Infrastructure, Subcommittee on Aviation.

Background

Overview

UAS have been a part of American aviation for nearly a century, primarily in military research and operations.[1] The FAA first authorized UAS operations in U.S. airspace in 1990.[2] Most operations since that time have been confined to public uses such as law enforcement and scientific research.

Recent and rapid advancements in computing technology have transformed UAS. Small unmanned aircraft are typically substantially less expensive, easier to acquire and simpler to operate than manned aircraft.[3] In some cases, UAS offer capabilities that cannot be matched by manned aircraft such as close inspections of oil rig flare stacks and transportation infrastructure.

As a result, there is tremendous demand for UAS as new applications are developed across industries beyond aviation including agriculture, energy and media. The economic opportunities and impacts are expected to be substantial. The Association of Unmanned Vehicle Systems International foresees that the use of UAS will lead to the creation of 100,000 jobs and $82 billion in economic impact by 2025.[4]

Governing Law and Regulation

In the *FAA Modernization and Reform Act of 2012* (P.L.112-95), Congress directs the FAA to take steps in furtherance of UAS integration into the National Airspace System.[5] This legislation directs the FAA to create a comprehensive plan for the integration of civil UAS by September 2015 and to issue regulations applicable to the operation of small UAS.[6] The FAA published a proposed rule for the Operation and Certification of Small UAS in February 2015, and reports that the final rule will be completed sometime in 2016.

Section 333 of the same legislation directs the FAA to permit UAS operations meeting certain criteria prior to the completion of the required comprehensive plan and rulemaking. After some delay, the FAA began permitting operations under section 333 in November 2014 for a variety of commercial applications, including surveying, photography and pipeline inspection. As of September 24, 2015, the FAA had granted 1,732 permits under section 333.

Section 336 defines "model aircraft" as any unmanned aircraft that are capable of sustained flight, flown within visual line of sight of the person operating the aircraft, and flown for hobby or recreational purposes. Section

336 prohibits FAA from issuing any rule or regulation regarding model aircraft, provided:

1. the aircraft is operated in accordance with a community based set of safety guidelines and within the programming of a nationwide community-based organization;
2. the aircraft is limited to less than 55 pounds unless otherwise certified;
3. the aircraft is operated in a manner that does not interfere with and gives way to any manned aircraft; and
4. notice is given to an airport operator or airport air traffic control tower if flown within 5 miles of an airport.

If an operator of an unlawful UAS operation is identified, the FAA and other agencies can pursue civil penalties for careless and reckless operations or criminal penalties for other legal violations.[7] To date, the FAA has initiated 20 legal enforcement actions related to unlawful UAS operations. The FAA is also working with state and local law enforcement agencies to address unlawful UAS operations.[8]

National Airspace System Safety

The proliferation of UAS in the United States has occurred at a dramatic rate, particularly as they have become widely available in the consumer market. In 2010, the FAA estimated that 15,000 UAS would be in operation by 2020.[9] Today, it appears sales of UAS exceed 15,000 each month and up to one million units may be sold during the 2015 holiday season.[10] The available data does not distinguish between those that will be purchased or sold as "model aircraft" used by hobbyists or other UAS used for personal or commercial purposes.

The rapid growth in unmanned aircraft (recreational, commercial and public) operated in the United States has been accompanied by a sharp rise in reported sightings by pilots of manned aircraft and air traffic controllers. Since last year, these sightings have been reported as occurring in proximity to major airports and during critical phases of flight.[11] The trend has continued into 2015 as pilots continue to report sightings to the FAA.[12] In August 2015, in response to a Freedom of Information Act request, the FAA released a compilation of over seven hundred reported sightings of UAS by pilots between November 2014 and early-August 2015. In addition, efforts to combat forest fires have been seriously disrupted as tanker crews have been forced to cancel or postpone missions because of unmanned aircraft sightings[13] In 2015, the U.S. Forest Service reported 18 unauthorized unmanned aircraft incursions above or near wildfires; 10 of these events hampered the Forest Service's aviation operations.[14] In the majority of cases, authorities have been unable to ascertain the identity of the unmanned aircraft operators. In addition to reports by pilots, unmanned aircraft have posed safety risks on the ground. In the last year, unmanned aircraft have crashed into the White House grounds, the stadium at the U.S. Open and a college football game.

There are questions about the nature of some of the sightings reported to the FAA.[15] For instance, questions have been raised as to whether certain reported sightings involve birds rather than unmanned aircraft or were instances in which an unmanned aircraft was visible to pilots but did not entail a loss of safe separation between aircraft.[16] In one case, a collision between an airliner and a UAS reported by a flight crew was later determined to be a bird strike.[17] In other cases, some questionable unmanned aircraft operations included in the FAA data were conducted by government agencies (including a local police department and the military) rather than errant consumers.[18]

The increase in reported UAS sightings by pilots as well as accelerating sales of unmanned aircraft to consumers concern stakeholders and the public for several reasons. First, safety is paramount, and unauthorized operations of unmanned aircraft create the possibility of a midair collision with a conventional aircraft that could result in injuries or the loss of life. Furthermore, other stakeholders say the FAA should develop methods for objectively measuring and analyzing the risks associated with UAS operations. Finally, stakeholders have expressed concern that unnecessarily restrictive government policies or the actions of irresponsible individuals could hamper or even cripple the development of the commercial UAS industry.

Solutions

Efforts are underway to address these risks through improved education, better enforcement of existing laws and technological means. For example, FAA has joined industry trade groups, manufacturers and other stakeholders on "Know Before You Fly" an educational campaign to provide prospective users of unmanned aircraft with the information they need to safely and legally operate in the nation's airspace. There are also private sector-led efforts to use technologies such as geo-fencing and enhanced aircraft detection to mitigate the risks.

This hearing will enable the Subcommittee to better understand the nature and extent of potential risks posed by unmanned aircraft to aviation safety, how to measure and analyze possible risks, and what technological, educational and policy solutions should be used to mitigate such risks.

End Notes

[1] John David Blom, Unmanned Aerial Systems: A Historical Perspective, Occasional Paper 37, pp 46. Combat Studies Institute Press, US Army Combined Arms Center. Available at: http://usacac.army.mil/cac2/cgsc/carl/download/csipubs/OP37.pdf.

[2] http://www.faa.gov/uas/faq/media/1009 UAS Fact Sheet.pdf.

[3] Small unmanned aircraft are those weighing less than 55 pounds.

[4] The Economic Impact of Unmanned Aircraft Systems in the United States, AUVSI, Mar. 2013. See: http://www.auvsi.org/auvsiresources/economicreport.

[5] Pub. L. 112-95, 126 Stat 11. (Feb. 14, 2012).

[6] Id.at 126 Stat. 73, Sec. 332.

[7] See 49 U.S.C. §§ 44709, 46301. In addition, chapter 463 of title 49 authorizes fines and/or imprisonment under title 18 for violations of national defense airspace, interference with air navigation, and transportation of hazardous materials. 49 U.S.C. §§ 46307-46317.

[8] FAA Issues UAS Guidance for Law Enforcement (Jan. 2, 2015) available at https://www.faa.gov/news/updates/?newsId=81244.

[9] Fed. Aviation Admin. FAA Aerospace Forecast Fiscal Years 2010-2030, Unmanned Aircraft Systems, available at: https://www.faa.gov/data_research/aviation/aerospace_forecasts/media/2010%20Forecast%20Doc.pdf Pg. 48.

[10] Unmanned aerial vehicles, Welcome to the Drone Age, The Economist, Sept. 26, 2015. & Aaron Karp. FAA warns of 'a million drones under people's Christmas trees', Air Transport World, Sept. 28, 2015. Available at: http://atwonline.com/technology/faa-warns-million-drones-under-people-s-christmas-trees.

[11] Craig Whitlock. Near-collisions between drones, airliners surge, new FAA reports show, Wash. Post, Nov.26, 2014. https://www.washingtonpost.com/world/national-security/near-collisions-between-drones-airliners-surgenew-faa-reports-show/2014/11/26/9a8c1716-758c-11e4-bd1b-03009bd3e984_story.html.

[12] Craig Whitlock. FAA records detail hundreds of close calls between airplanes and drones, Wash. Post. Aug. 20, 2015. https://www.washingtonpost.com/world/national-security/faa-records-detail-hundreds-of-close-calls-betweenairplanes-and-drones/2015/08/20/5ef812ae-4737-11e5-846d-02792f854297 story.html.

[13] Joseph Serna, Drone sighting halts some firefighting efforts in Northern California, Los Angeles Times, Sept. 14, 2015.

[14] Information provided to Subcommittee staff.

[15] See A Closer Look at the FAA's Drone Data, available at http://www.modelaircraft.org/gov/docs/AMAAnalysisCloser-Look-at-FAA-Drone-Data_091415.pdf.

[16] Bart Jansen. Drone hobbyists find flaws in "close call" reports to FAA from other aircraft, USA Today, Sept. 14, 2015.

[17] Whitlock 2015. Supra.

[18] Jansen, Supra.

In: Unmanned Aerial Systems
Editor: Wilfred Sparks
ISBN: 978-1-63484-652-3

Chapter 7

STATEMENT OF MICHAEL G. WHITAKER, DEPUTY ADMINISTRATOR, FEDERAL AVIATION ADMINISTRATION. HEARING ON "ENSURING AVIATION SAFETY IN THE ERA OF UNMANNED AIRCRAFT SYSTEMS"*

Chairman LoBiondo, Congressman Larsen,
Members of the Subcommittee:

I am pleased to appear before you today to discuss a subject that continues to be the topic of a lot of conversation; Unmanned Aircraft Systems or UAS. Wherever we look, in everything from popular culture to store shelves, UAS seem to be everywhere. It might appear to some people that UAS suddenly appeared in our skies and are now everywhere, from the White House lawn to the U.S. Open Tennis tournament, sometimes flying too close to commercial aircraft or interfering with firefighting efforts. Many different departments and agencies within the federal government have responsibilities associated with UAS. Among other things, the Federal Aviation Administration's (FAA) responsibility includes the safe and efficient integration of UAS into the National Airspace System (NAS). In 2012, Congress passed the FAA

* This is an edited, reformatted and augmented version of a statement presented October 7, 2015 before the House Committee on Transportation and Infrastructure, Subcommittee on Aviation.

Modernization and Reform Act of 2012 (2012 Act), which, in part, charged the FAA with safely integrating UAS into the NAS by September 30, 2015.

FAA has recognized the significance of this technology and has adapted organizationally to provide this emerging technology with a commensurate level of attention. Recently, FAA selected two executives to oversee our UAS integration efforts. Accompanying me today are Marke "Hoot" Gibson and Earl Lawrence. Hoot is the Senior Advisor on UAS integration, reporting directly to me. He will establish a focus on external outreach and education and interagency initiatives. Earl is the Director of the UAS Integration Office within the FAA's Safety organization. He will lead the FAA's efforts to safely and effectively integrate UAS into the NAS. Their addition to the FAA team acknowledges the expanded demand of UAS issues both inside and outside the FAA.

There is no doubt that UAS can be of great value to this country. In accordance with the appropriate authorizations, UAS are being used today to examine infrastructure, survey agriculture, provide emergency response support, examine damage caused by time or disaster, and go places that would otherwise be dangerous for people or other vehicles. Entrepreneurs around the world are exploring innovative ways to incorporate the potential of UAS into their corporate activities. There are a number of public and commercial operations being conducted today, including the ones mentioned above, that contribute to public safety and enhance the ability of corporations to achieve important goals. FAA does not underestimate the importance of integrating the range of UAS technology into the NAS, but there are significant safety challenges that must be mitigated for this to occur.

For example, we have witnessed a huge influx of casual users, people who fly UAS for entertainment or recreation. This has become the crux of a growing problem. UAS introduces, not just a new class of aircraft, but a new class of pilot. The vast majority of these operators do not have the basic aviation training or experience required for pilots of traditional aircraft. They have no knowledge that they may be flying in controlled airspace. Some may have no recognition that their actions could have serious consequences. They are simply having fun with a toy.

The primary goal of the FAA is to integrate this new class of aircraft and their operators safely and efficiently into the NAS, regardless of whether the operations are recreational or commercial in nature. Because this new branch of aviation is changing at the pace of human imagination, the FAA believes a flexible framework is imperative. The UAS industry is developing many new exciting technologies and the FAA must provide a regulatory framework for

UAS to operate safely. Our goal is to provide the basic rules for operators, not identify specific technological solutions that could quickly become outdated. The FAA is creating a safe operational environment for innovators to demonstrate their technologies. We are doing this through the establishment of basic operational regulations, the issuance of exemptions and experimental certificates, and through our continued research and collaboration at our UAS test sites, Center for Excellence, and Pathfinder programs.

Our efforts also include long-term planning, including the ongoing development and finalization of the regulation of small unmanned aircraft. We are conducting collaborative research and development with interagency partners and the UAS industry. We have established test sites and airspace for these activities. The FAA and our government partners have always realized that the best way to succeed is through partnership, whether it be with industry or other governments. Finding consensus leads to cooperation and willing participation.

Consistent with our approach to other regulations, we are establishing a risk-based approach to the regulations in this area, laying a strong foundation for safe integration. The concept is balance. We must develop a broadly scoped approach to rulemaking to identify and mitigate safety risks without stifling innovation and industry performance. However, a key factor in the success of any regulation is the willingness of the operators to follow them.

Integrating UAS means integrating operators into the aviation culture and mindset. It means creating a general awareness that these devices are not toys and the consequences of misuse can be serious. We believe the most effective way to accomplish this task is through education. We want to work through partnerships with model aircraft organizations, manufacturers, and interagency partners, as well through traditional and social media outreach, to ensure that these new operators know when and where they may safely fly.

To this end, with the help of our stakeholders, we developed the "Know Before You Fly" and "No Drone Zone" programs. "Know Before You Fly" offers common sense advice, such as don't fly near airports, don't fly in adverse weather, don't fly under the influence of alcohol or drugs, and don't fly over people or sensitive infrastructure like power plants.

Our "No Drone Zone" program began with the Super Bowl earlier this year and is tailored to specific events and places, such as the heavily restricted airspace around Washington, D.C. The No Drone Zone video posted on YouTube prior to the Super Bowl this year received over 59,000 hits. Most importantly, we received no reports of unauthorized activity in the restricted airspace around the stadium.

We want people to enjoy their hobby, but we want to make sure they fly safely. Education, such as the programs noted above, has been our preferred method for successfully integrating UAS operators. We can never let an educational opportunity slip by. We need to be creative and collaborative in our approach to reaching the public. I will share with you two examples of this approach. Several UAS manufacturers have started to voluntarily include the "Know Before You Fly" safety literature with their product packaging. We are also trying to reach a broader audience by working with the San Francisco 49ers NFL football team to use their scoreboard to make public service announcements during their games. We hope to expand this type of outreach in the coming months.

In order to make it easier for the operator to know when and where it is appropriate to operate a UAS, in particular a model aircraft, we helped develop the B4UFLY mobile app. This is a simple, intuitive user interface that lets the UAS operator know if there are any prohibitions in place for where they are flying or where they want to fly. There is a color-coded status indicator with text that provides the operator with situational awareness. Beta-testing of this app is ongoing and the FAA will make adjustments where necessary based on user feedback. The industry is clearly looking to FAA for leadership in educating the public about the safety parameters for model aircraft operations. Our vision for this app is that the FAA would continue to support the basic technology, while other companies could augment it with their navigational maps.

As discussed, the FAA believes that partnerships and education are the keys to the success of safe UAS integration. But to be clear, if the unauthorized operation is intentional or is intended to cause harm, strong and swift enforcement action, including criminal enforcement, will be taken.

When UAS delayed fire-fighting activities in the drought-stricken western states, local law enforcement and forest service personnel were on the front lines dealing with the situation. We are working with law enforcement agencies to educate them about our rules and to emphasize that, in addition to the FAA's rules, there are existing state and local laws in areas of reckless endangerment, trespass, and privacy that could apply. Just because this is a new technology or different than what law enforcement has seen in these areas before does not mean that these laws would not be equally applicable to such acts involving use of UAS. We want to work with law enforcement because if they encounter unauthorized UAS operations, they can help us to gather evidence and find witnesses that will help with our investigations and enforcement action. For some, education will never be sufficient. As with any other activity, we will always have to contend with those who wish to cause mischief or refuse to consider the potential harm their activities might pose to others.

Informing and educating UAS operators is just one piece of integrating these vehicles safely into the NAS. The 2012 Act provided the Secretary with the authority to issue exemptions that allow for commercial UAS activity in low risk, controlled environments (section 333 exemptions). After gaining experience with various types of operators, the Department recently expedited its approach for section 333 exemptions. We are now able to issue summary grants when we find that we've already granted a similar exemption. Summary grants are more efficient because they do not require applicants to repeat analysis that has already been performed. This streamlined approach now allows the Department to issue between 40 and 50 section 333 exemptions a week. These exemptions are effectively acting as a bridge until the small UAS rule comes out to more broadly authorize the operation of UAS under 55 pounds if the operations take place under a set of parameters to maintain safety, including operating at speeds below 100 mph and below 500 feet in altitude.

Additionally, as part of our efforts to streamline the integration of this technology, the FAA has further expedited safe UAS integration and facilitated commercial use by issuing a blanket Certificate of Waiver or Authorization (COA) for flights at or below 200 feet when it issues a section 333 exemption. The blanket COA can be used for UAS operations that involve aircraft that weigh less than 55 pounds, operate during daytime Visual Flight Rules (VFR) conditions, operate within visual line of sight (VLOS) of the pilots, and stay at the prescribed distances away from airports or heliports.

From the outset, we have worked closely and successfully with government partners and industry stakeholders to achieve milestones put forward by the 2012 Act. In coordination with other governmental agencies and industry, we developed two long-term planning documents, the Comprehensive Plan and a five-year Roadmap. We have worked with members of the UAS Executive Committee (ExCom), comprised of representatives of various government agencies and departments with responsibilities in this area, to leverage our collective assets and conduct research and development on UAS integration while ensuring the continued safety of the NAS. The FAA collaborated with the National Aeronautics and Space Administration (NASA) on studies advancing air traffic control interoperability with future use by UAS of detect-and-avoid (DAA) systems in controlled airspace. We continue to collaborate with the industry on flight tests to validate RTCA[1] standards for DAA systems as well as command and control radios. RTCA began work on the standards at the request of the FAA in 2013 and they are scheduled for completion in 2016. These standards will help resolve two of the difficult challenges facing the industry for integration of UAS into the NAS. NASA, the FAA, and industry partners have successfully demonstrated a proof-of-concept airborne DAA system and prototype radios for use as command and control systems for UAS.

We are already looking beyond the small UAS rulemaking at what comes next in terms of the types of operations expected, and what technologies we may need to certify to ensure safety. The FAA has consulted with the UAS ARC to determine the next areas of focus so we can enable those UAS operations with the highest net societal benefits. These recommendations are being assessed and will result in additional focus areas that will become the centerpiece for FAA's strategic plans for UAS integration.

As the aerospace industry and aviation system grow more complex, we must ensure that our resources are directed to those areas which pose the greatest risk to safe aviation operations. We will need to expand collaborative, data-driven processes with the UAS industry to improve safety and streamline process in areas such as certification. We must meet challenges and take advantage of opportunities.

The safe integration of UAS into the NAS will be facilitated by new technologies being deployed as part of the Next Generation Air Transportation System (NextGen). NAS Voice System (NVS), Data Communications (Data Comm), and System Wide Information Management (SWIM) will provide more information, flexibility, situational awareness, and a greater ability to communicate with NAS users.

To enhance safe application of new and emerging technologies, earlier this year, FAA established the Pathfinder Program, which was referenced above. In the Pathfinder Program we work with three companies to obtain important information on the next steps beyond operational parameters included in the small UAS notice of proposed rulemaking. For visual line of sight operations, FAA is working with CNN on how UAS might be used for news gathering in populated areas. We are also working with the UAS manufacturer Precision Hawk to explore beyond visual line of sight operations in rural areas. Precision Hawk will be working to explore how flying beyond the pilot's direct vision might be used to allow for greater UAS use for crop monitoring in precision agricultural operations. BNSF Railway will explore command and control challenges of using UAS to inspect rail system infrastructure. Developing the safe use of this important technology can only benefit how UAS can be used in the future.

Aviation technology is constantly evolving. This is certainly not the first time we, as an agency, have been required to integrate new aviation technology into the NAS. Different aircraft technologies, including jet engines, were required to be accepted operationally and we handled them as they developed. Today, in addition to UAS, we are working to integrate commercial space technology into the NAS. Clearly, there will be other technologies that we will be required to integrate moving forward.

I am proud of the team we have brought together and of the approach we are taking to ensure that our airspace continues to be the safest in the world, even as we work to accommodate new technologies that have the potential for changing the way we live our lives. This is an exciting time to be part of the FAA. I am happy to have had the opportunity to speak with you this morning and I will be glad to answer any questions you have.

End Note

[1] RTCA, Inc. is not-for-profit organization that serves as a federal advisory committee to the FAA. See http://www.RTCA.org.

In: Unmanned Aerial Systems
Editor: Wilfred Sparks
ISBN: 978-1-63484-652-3

Chapter 8

STATEMENT OF JAMES HUBBARD, DEPUTY CHIEF, UNITED STATES FOREST SERVICE. HEARING ON "ENSURING AVIATION SAFETY IN THE ERA OF UNMANNED AIRCRAFT SYSTEMS"*

Mr. Chairman and Members of the Subcommittee, thank you for the opportunity to present the views of the Forest Service regarding the safety of Unmanned Aircraft Systems. Increasingly severe fire seasons are one of the greatest challenges facing the Nation's forests and 2015 has been no exception. The Forest Service and the Department of Interior (DOI), in full coordination with our state and local fire response partners, are responsible for responding to unplanned wildland fire incidents which threaten lives, property and resources. The agency mobilizes ground crews, fire engines, helicopters, airtankers and other assets in our suppression efforts. At the same time we are investigating the potential of using Unmanned Aircraft Systems (UAS), we have also experienced unwanted incursions from recreational drones in our suppression efforts.

* This is an edited, reformatted and augmented version of a statement presented October 7, 2015 before the House Committee on Transportation and Infrastructure, Subcommittee on Aviation.

Wildland Fire Suppression

The United States has developed a complex and sophisticated wildland fire response framework that is composed of local, state, tribal and federal entities. Federal and state land and fire managers, Tribes, non-governmental organizations, and other stakeholders work as partners to safely and effectively extinguish fire when needed; use fire where allowable; manage our natural resources; and as a Nation, live with wildland fire. Each entity has specific responsibilities for responding to wildfires occurring within their jurisdiction (initial response). Federal, state and local entities also coordinate and share resources and responsibilities as fires become larger and exceed the local response capacity. Fortunately, local response capacity is generally all that is needed for controlling or extinguishing most wildfires, though escalating risks including more people moving into (and next to) wildland areas may impact response effectiveness in the future.

During this fire season, we spent 24 days at National Preparedness Level 5 -- the highest level -- meaning all available ground and air assets are committed to priority work. Severe drought across the west has increased fire severity in several states. Washington State, among others, recorded a record season of severe wildfires. Drought-ridden California also experienced tremendous fire activity. The Forest Service and DOI, in coordination with our state and local fire response partners, mobilized over 27,000 firefighters along with numerous airtankers, helicopters, fire engines and other assets through our integrated, interagency suppression efforts.

Every state, Puerto Rico and Pacific Island, along with the military and international support, has provided people and equipment this season to help manage over 50,000 wildfires.

This season, over 9 million acres burned across the United States and destroyed over 2,500 single residences. The greatest losses, however, involved the fatalities of 13 wildland firefighters who made the ultimate sacrifice to protect the lives of others. Our thoughts and prayers will forever be with their family and friends.

Airspace Above Wildland Fires

Aviation operations are one of the highest risk suppression activities we perform. Large airtankers and helicopters are often mobilized during initial response and extended attack suppression efforts. Often, we request and receive a temporary flight restriction and a Notice to Airmen to maximize safe airspace over wildfires. The airspace above large fires is very complex. Firefighting agencies have established a communication protocol known as the Fire Traffic Area (FTA) to reduce risk in the airspace over a wildfire. The FTA establishes a protocol for communication and operations by firefighting aircraft. In general, fire aircraft must make radio contact with the command and control "Air Attack" when they are 12 nautical miles from the fire. If contact is not made, the fire aircraft will remain outside the area orbiting a minimum of seven miles away. Once cleared into the FTA, there is a structure based on altitude for the various missions. Air Attack remains at higher elevations above the tactical aircraft to retain the best visual perspective. Airtankers waiting their turn stay higher in altitude than those maneuvering for a retardant drop. Helicopters operate at the lower altitudes as their mission often involve transport of firefighters, cargo or water. The complexity of the airspace above a wildfire can rapidly change as fire activity increases, resources arrive on scene and smoke reduces visibility. Safety depends on understanding what other aircraft are in the airspace and where those aircraft are operating.

USE OF UNMANNED AIRCRAFT SYSTEMS

UAS are among the technologies that hold significant potential to aid in fire detection and increase firefighter safety efforts. We aim to be a leader in embracing the new technology and enhancing our capacity to accomplish our mission to "sustain the health, diversity, and productivity of the Nation's forests and grasslands to meet the needs of present and future generations." The Forest Service established a UAS Executive Steering Committee to coordinate efforts with our state and federal partners. The Executive Steering Committee is charged with developing policy regarding the operation and use of UAS, including natural resource research along with recreational use on National Forest System (NFS) lands. Through Executive Steering Committee guidance, policy will be evaluated and updated with UAS technology in mind to ensure safe and efficient integration of UAS in government mission and public safety for forest visitors.

As a part of our efforts to explore new technologies to enhance firefighter safety and effectiveness, we partnered with DOI's Office of Aviation Services to develop a 2015 Unmanned Aircraft Technology Demonstration Strategy.

The Strategy outlines a five-phase process enabling us to test UAS capabilities to support fire suppression efforts by completing scenario-based missions. The Forest Service and DOI will develop summary of the activities during the 2015 field season, along with lessons learned, and will then make recommendations on further integration of UAS on a national level.

Although the focus of today's hearing is UAS and wildfire, the Forest Service is aware of the potential of UAS to augment the agency's capacity to gather information to support several resource management programs. Examples include mapping, monitoring the condition of natural resources, assessing the effectiveness of natural resource management projects, engineering inspections, and public use inventories.

We have identified five main ways the wildland community can benefit from UAS in fire management including:

1. *Increasing awareness of ground resources regarding real time fire characteristics, such as location, behavior and spread.* A small UAS above the tree canopy can help ground assets learn more about fire activity in their vicinity. In some situations, heat signatures or infrared sensors could help in smoky or other low visibility situations.
2. *Enhancing safety through scouting fire line and identifying escape routes and safety zones.* Significant potential exists to enhance firefighter safety by aerially assessing the terrain, condition of fuels, fire behavior, and potential escape routes without putting a human in dangerous situations.
3. *Tracking and monitoring ground resources in real time.* UAS technology can help provide accurate locations of personnel and equipment and direct them to strategic locations or advise them out of dangerous situations.
4. *Enhancing communication links.* Use of UAS can help link radio communications when terrain hinders line of sight transmissions or over a larger area for longer term incidents.
5. *Supporting aerial supervision and mapping in low visibility situations.* UAS tools can observe ground crews and providing mapping capabilities in smoky or other low visibility situations.

We continue to investigate the future possibilities for potentially deploying UAS for direct attack and supply support.

DRONE INCURSIONS

In 2014, we experienced four incidents during the fire season where a non-participating UAS was flown over or near a wildfire, neither of which affected firefighting aircraft.

This season, there have been 21 documented instances of UAS incursions near and over wildfires with ten being over NFS lands. Twelve of the incidents required firefighting aircraft to stop operations and shut down, relocate to another area or be grounded at base. In two of the events, firefighting pilots reported taking aggressive maneuvering action to avoid a collision. Approximately five hours of operational time was lost while ensuring the risk of mid-air collisions had been mitigated. These incursions by untrained operators inhibited the ability to deliver retardant and water to slow the fire's advance; prevented transportation of firefighters to vital locations for fire suppression and containment efforts; and stopped cargo delivery.

While it may be hard to quantify how shutting down air operations affects a wildfire's growth, we do know a rapid response including aviation assets, is successful in suppressing fires while they are small and controllable. In a well-publicized UAS wildfire incursion in Southern California, five non-participating UAS were flown during critical fire suppression activities on the North Fire. At the time the first non-participating UAS was spotted, the fire was only 4-5 acres, but was spreading rapidly. The two airtankers over the fire were directed by a smaller fixed-wing aircraft known as the lead plane. Two helicopters were dropping water in an effort to slow the fire's growth. Once a UAS was observed in the airspace over the fire, all six firefighting aircraft broke off suppression activity and relocated away from the fire. The command and control aircraft, known as air attack, then proceeded to assess the airspace.

Multiple UAS were identified over the fire. After an approximate 20-25 minute delay, the firefighting aircraft re-engaged with suppression activities. By then, the fire had crossed over Interstate 15, with life and property at risk. The North Fire ultimately consumed 4,250 acres, burned or destroyed 74 vehicles, and destroyed 23 structures before it was stopped. Fortunately, there were no injuries reported.

Coordination with Other Federal, State and Local Agencies

The Forest Service is coordinating with other federal, state and local agencies on several efforts including utilizing UAS in the National Airspace System, public education campaigns and UAS monitoring technology.

Utilizing UAS in the National Airspace System

The Federal Aviation Administration (FAA) and DOI have a Memorandum of Agreement (MOA) to allow DOI to fly UAS in the National Airspace System under a notification system for over a year. The Forest Service is currently working toward an MOA with FAA as well. The MOAs streamline airspace approval and reduce processing time to obtain a Certificate of Authorization. The Forest Service is grateful that the FAA has been active in helping to find solutions to the UAS wildfire incursions and in facilitating UAS integration into Forest Service operations.

Public Education Campaign

In 2014, shortly after the first UAS was seen near a wildfire, the Forest Service, in partnership with other federal, state, and local agencies issued a national news release urging members of the public not to fly UAS near wildfires. Shortly thereafter, the Forest Service developed a poster for use in public education and outreach with the slogan "If you fly, we can't". The campaign has evolved to include a series of tools to help promote aviation safety including several public service announcements and short videos, news releases, and Facebook and Twitter posts.

As UAS incursions began to increase, a concerted public education effort began through news releases, social media posts, local news conferences, preliminary discussions with the Ad Council, direct outreach to the UAS industry and hobbyist communities through the FAA "Know before you fly campaign," the Association for Unmanned Vehicle Systems International, and the Academy of Model Aeronautics. We continue to coordinate with the FAA, and DOI to identify and coordinate public education efforts and determine success can be accomplished.

UAS Monitoring Technology

The United States Department of Agriculture (USDA) Office of Homeland Security and Emergency Coordination (OHSEC) facilitated UAS technology coordination between the Forest Service and the Department of Homeland Security (DHS). DHS is adding wildfires to their modeling process to help identify technology that might be of benefit to fire managers in addressing UAS incursions. The Forest Service is providing DHS with information about fire operations, fire environment and non-participating UAS airspace issues and will provide opportunities to assess technologies in live fire environments as the project progresses.

CONCLUSION

The use of UAS holds significant potential to increase firefighter safety, and to assist in suppression efforts. At the same time, we remain very concerned about the dangers to firefighters and the public associated with non-participating UAS incursions in airspace above a wildland fire. We look forward to working with the subcommittee to identify specific protections that would allow wildland firefighters to take action on individuals who operate UAS over emergency situations.

In: Unmanned Aerial Systems
Editor: Wilfred Sparks

ISBN: 978-1-63484-652-3

Chapter 9

STATEMENT OF CAPTAIN TIM CANOLL, PRESIDENT, AIR LINE PILOTS ASSOCIATION. HEARING ON "ENSURING AVIATION SAFETY IN THE ERA OF UNMANNED AIRCRAFT SYSTEMS"*

The Air Line Pilots Association, International (ALFA) represents more than 52,000 pilots who fly for 31 airlines in the United States and Canada. Thank you for the opportunity to provide our perspective on the critical importance of safely integrating unmanned aircraft systems (UAS) into the U,S, national airspace system (NAS), the most dynamic and diverse such system in the world. The NAS must be protected and maintained to deliver the safest and most efficient air transportation services in the world, Although our focus today is the NAS, we must point out that the safety issues highlighted are independent of any national airspace boundary and are faced by ALPA's pilots as we operate around the globe.

* This is an edited, reformatted and augmented version of a statement presented October 7, 2015 before the House Committee on Transportation and Infrastructure, Subcommittee on Aviation.

UAS Risk Must Be Effectively Managed to Realize Benefits

ALFA recognizes that UAS represent a significant potential for economic and societal benefit, They are uniquely suited for performing many types of dangerous flying that can keep pilots out of harm's way, ALPA supports robust development of this technology with one single overriding condition: integration of UAS into the NM must be done safely, deliberately, thoughtfully, with full understanding of the possible risks also being introduced, and most importantly—with simultaneous development of effective mitigations for those risks. We have to do this right, or the enviable safety record we have achieved in airline operations will be at risk, and with it, the promise of employing unmanned systems for the benefit of the population.

As we have for many years, ALPA continues to be an active partner with both government and industry in developing standards that will lead to safe operation of unmanned systems in the NAS. Concurrently, we recognize that these standards are far from complete. Defining a safety framework for any new technology is necessarily a painstaking process, and ALFA, along with hundreds of extremely talented representatives from across aviation, is diligently pursuing that goal.

UAS Sales Growth

Among the most dramatic and challenging revolutions in aviation technology and operational capability to be introduced Into the NAS is the UAS. There are many different types of UAS, and there seems to be an ever-expanding list of potential applications.

There are many published data sources that support this key fact: *there is tremendous growth in the sale and use of small UAS for both hobby and commercial use.* Research suggests that there are likely more than 700,000 UAS already in the hands of the public, many of whom undoubtedly have very little appreciation for, or understanding of, how to safely operate UAS in airspace that is used by commercial aviation.

Additionally, the FAA has stated that upwards of 1 *million* UAS could be sold this coming holiday season. Although the FAA has made strides toward developing a regulatory framework for the operation of some of these aircraft, much work remains. In the meantime, the FAA is publishing data that increasingly point to a UAS threat to safe operation of airline aircraft. Immediate action is required to address the documented hazards.

FAA Addresses Commercial Small UAS Operations

The FAA has taken initial steps to allow small UAS (sUAS, under 55 pounds) to begin operating in the airspace system with multiple restrictions intended to mitigate risk. The FAA has established an interim process of approval of commercial operations on a case-bycase basis. This is often referred to as the "Section 333 process" because the FAA's use of this strategy is based on that section of the most recent FAA authorization.

In addition to the interim approval process, the FAA published a notice of proposed rulemaking (NPRM) on Feb, 23, 2015, that addressed the commercial operations of sUAS. The NPRM, which was significantly based on FAA's 2009 recommendations of the Small UAS Aviation Rulemaking Committee, established a proposed framework for commercial operators to operate their sUAS. The NPRM formally established the definition of an sUAS, established pilot qualifications, and created operational limitations.

By the end of the notice's 60-day comment period, the FAA had received more than 4,000 comments. While ALPA supports the FAA's efforts to properly establish regulations for commercial sUAS operations, ALFA and other organizations encouraged the FAA to make some substantial changes.

The FAA is now reconciling ALPA's comments along with all of the others and will ultimately issue a final rule. Between the Section 333 process and the eventual sUAS rule, commercial operators are well on their way to having a defined path for approved NAS operations and a path for expansion of operations while insuring safety.

The FAA Needs to Address All UAS to Ensure Safety Immediately

The standards for some of the key capabilities of UAS, and the recommendations for the wide variety of rules that must be changed or developed to accommodate large UAS, are still years away. As a result, for the foreseeable future, without additional FAA action there will be no rules for the following UAS operations:

- Non-commercial operations by companies (e.g., pipeline or power line patrol by the pipeline/power line company employees)
- Recreational/hobbyist small and large UAS
- Commercial large UAS

ALPA recognizes that the commercial operations of large UAS are not developing as quickly as sUAS used for commercial, recreational and corporate operations. The FAA presently has rulemaking underway that addresses only one of these four types of UAS operations. The tremendous growth of sUAS in just the last 24 months when measured against the limited rules that the FAA has under way is inconsistent with the needs of the country for safe integration of UAS.

RECENT FAA INCIDENT REPORT DATA IS DEMONSTRATIVE OF ALFA SAFETY CONCERNS

This August, the FAA published a list of pilot reports on UAS encounters. ALPA reviewed the 764 events, which cover only a 10-month period from November 2014 through August of 2015. Both the volume of events and many of the event descriptions are sobering reminders to the industry that the risk of a collision between a UAS and an airline aircraft has increased significantly. Consider these sample summaries (ALFA paraphrase):

New Jersey:

- 07/16/2015 Summary: Regional turboprop aircraft on a one-mile final approach to Runway 11 reported a remote controlled aircraft 50 feet below at Weequahic Park.
- 08/09/2015 Summary: Several aircraft on final to Newark Liberty International Airport reporting unidentified drone operating directly on final approach to Runway 4 right. The drone was reported to be at altitudes between 2,000 and 3,000 feet, between 8 miles and 13 miles out on the final approach. Three aircraft reported the drone in sight.

Oregon

- 05/17/2015 Summary: Regional jet at 1,000 feet on a two-mile final for Runway 34 left. Pilot reported seeing a drone at 600 to 700 feet off his left wing, It was approximately 400 to 500 feet from the aircraft. The drone was described as a yellow hobby-style helicopter with four rotors the size of a large bird.
- 07/21/2015 Summary: Turboprop commercial freight aircraft reported a "drone' at 1,100 feet on base for Runway 28 left over 'Rocky Butte" (a hill approximately 2.5 miles southeast of Portland International Airport]

Washington

- 07/15/2015 Summary: Regional let reported a drone at 2,400 feet on a 5-mile final to Runway 16 right. The pilot reported that the drone was above him on final and appeared to be a black quadcopter.
- 04/27/2015 Summary: Jet airliner was descending through 3,500 feet on the downwind near Vashon Island (near Seattle) and sighted a deep blue metallic drone. The drone passed about 100 to200 feet under his left wing and appeared to be traveling northwest.

ALFA also studied the work of others who have reviewed the FAA event data, The Center for the Study of the Drone, which is operated by Bard College in New York State, has indicated that of the total number of reported events in the database, 302 of the 764 reports were determined to be near misses. In other words, its analysis revealed that 39.5% of the FAA's events would potentially have met the FAA's near mid air collision (NMAC) criteria, In comparison, ALFA research of FAA data regarding manned-on-manned aircraft NMAC finds that a total of 41 NMAC reports were logged in the same period. It should be noted that FAA NMAC reports are voluntarily submitted and therefore likely not a comprehensive set of manned-on-manned NMAC reports,

It is clear that hundreds of NMAC of UAS over such a short time frame far exceeds an acceptable level of risk to manned aircraft in the NAS. Undoubtedly there will be many perspectives and opinions on what constitutes an acceptable level of risk. If the FAA UAS event data tallied 100 instead of 764, ALFA would still insist that there are too many unexpected encounters.

Instead of discussing subjective opinions on the risk that UAS pose on manned aircraft, ALFA suggests that the FAA should invite ALFA and others in the industry to work collaboratively to reach an agreement on the level of risk that is deemed acceptable, and then work to implement solutions to achieve the targeted risk levels. The rate of UAS encounters needs to be reversed, and ALFA is ready and willing to immediately contribute in a meaningful way to reverse the growing trend of UAS encounters.

Non-Commercial and Recreational UAS Operations Appear to Be a Major Source of Reported UAS Events

Although the FAA has made progress in attempting to educate non-commercial and/or hobby users as to the safe operation of their aircraft through its "Know Before You Fly" campaign, no regulations exist that govern the operation of the aircraft or the training and experience of the pilots, ALFA believes that the vast majority of the many "close encounters" with unmanned aircraft reported by airline flight crews are the result of users who either do not understand the potential severity of operating near airports and aircraft, or are completely unaware that they are doing so. The massive growth of this segment clearly has outpaced the FAA's ability to effect safety standards that apply to it, yet the FAA remains responsible for the safety of all operations in the airspace. This is a significant gap that must be addressed. The FAA must have the ability to ensure the safety of the NAS regardless of the types of unmanned operations being conducted, and it must have the resources necessary to act on that mandate. As we have said before, we simply cannot afford to quantify this hazard by analyzing the damage after an unmanned system collides with an airliner.

UAS Frangibility Is Questionable and Untested

ALFA is concerned about the impact of sUAS on an airline aircraft if there were a collision. There are numerous videos of UAS crashes online; in many cases the crashes occur without major damage to the camera and the visible parts of the sUAS. It appears that the sUAS are generally designed to be relatively rugged, as one would expect for a commercially viable product. This ruggedness however, needs to be evaluated in the context of the potential damage that an sUAS would impose on an airline aircraft should the two collide. We frequently hear the comment that most sUAS are small, lightweight plastic aircraft. While this is the case for the sUAS airframe itself, the multiple propulsion units, batteries, and onboard cameras are hard metal with a significant density that a bird, for instance, does not have.

Airplane engines, windscreens, and other components may suffer the impact of such material without resulting in loss of the aircraft, *but* the damage would be nevertheless significant, jet engines, for example, are notoriously susceptible to foreign object damage (F0D) typically caused by small, hard objects found all over airports — nuts, bolts, rocks, tools, etc. Damage of this type, while rarely noted in conjunction with an accident, costs the industry billions of dollars every year.

Similarly, we have seen just in the past few months the extensive damage done to an aircraft in flight by hail, shattering both pilots' windscreens and severely damaging every part of the airplane that was hit. Here again, catastrophe was averted by the robustness of the airplane design and the skill of the flight crew, but the seriousness of striking hard objects in flight was clearly illustrated. Based on our experience with FOD, hail, and other objects striking transport aircraft, ALFA recommends design evaluations, modeling, and testing the collision impact of some of the more popular sUAS.

UAS Conspicuity - Data Suggests That They Are Difficult to See Until Close

ALPA is frequently asked to explain how visible an sUAS is to a flight crew of an airline aircraft There are medical studies about the limits of human visual acuity and some limited study data on distances at which a pilot perceives other aircraft. However, because UAS can be of essentially an infinite variety of sizes, shapes, and colors, studies regarding traditional aircraft do not provide a good guide. Factors such as size, shape, contrast with background, and movement relative to the observer are all factors that complicate a pilot's ability to see a UAS until it is extremely close and often too late to safely take evasive action. It is important to note that from a safety perspective, a pilot simply seeing an object in the airspace is only part of the process. The object must be seen with enough clarity and at such a distance that a pilot has the ability to identify the object and determine if evasive maneuvering is necessary to avoid a collision. To our knowledge, no specific quantifiable data on observing UAS from an aircraft moving nearly 200 mph in time to avoid collision exist.

ALPA Members Encouraged to Report UAS Sightings

In order to continue to track the rate of UAS encounters, ALPA has taken the initiative to encourage pilots who encounter an sUAS to follow their company guidance *and*, where applicable, file a report of the encounter with the FAA, A dedicated page on ALPA's website (www.alpa.org), available to anyone, not just members, helps pilots understand whether the encounter was likely reportable as a near mid air collision and provides web links and information needed to submit the necessary information.

True UAS Integration; Much to Do to Achieve One Level of Safety

While it is easy to focus on very near-term, dynamic growth and the recent news created by the FAA, it is important to understand the broader challenges that stilt exist for nationwide integration of UAS into the NAS. By way of review, UAS are grouped into two separate categories, generally based on weight. Aircraft that weigh 55 lbs. or less are defined as "small" UAS (sUAS) and unmanned aircraft that weigh more than 55 lbs. are considered "large."

Small UAS are typically intended for use at low altitudes, in clear weather, and within sight of the pilot. However, the aircraft themselves are capable of achieving altitudes, speeds, and ranges that extend far beyond the intended limits of their use. In addition, applications are being developed to employ these aircraft in an autonomous role, meaning they could operate along a predetermined flight path without being under the direct control of a pilot. While the commercial applications for sUAS are expanding rapidly, so too is the appeal to hobby users. In both *cases,* the potential exists for these aircraft to stray into airspace occupied by other NAS users, most notably by airline aircraft near airports.

Large UAS aircraft can be as large as a small airliner and can operate essentially anywhere conventional aircraft operate, and in many cases have performance capabilities exceeding those of conventional aircraft, While differences in size, performance, and operational capabilities can vary greatly, there also exists a wide range of technology on the ground that forms the entire system that must be considered in evaluating the safety of integrating these aircraft into the NAS, not just the aircraft itself.

Until comprehensive solutions are developed and promulgated by the FM, our overarching position is that no unmanned aircraft, public or civil, should be allowed unrestricted access to conduct flight operations into the NAS unless it meets all of the high standards currently required for every other airspace user. This means UAS must be designed to interoperate, with similar performance and functional requirements at the heart of their system, and have architectures embodying state-of-the-art safety technologies and system redundancies as required by currently certified commercial and general aviation airspace users. Of particular importance and concern is the ability of commercial aircraft operating in the NAS to safely perform see-and-avoid and collision avoidance maneuvers against UAS that may be operating in the same area.

ALPA's position is that the foundation of operating an aircraft, and the system functions therein, in a safe and responsible manner must be maintained at the same level of safety regardless of the location of the pilot or levels of automation. At the center of current commercial aviation flight operations is a well-trained, well-qualified professional pilot. A well-qualified pilot remains the single most important safety component of any commercial aircraft. A UAS should be able to operate as a part of the WAS, with similar certification standards and regulations to meet the target level of safety that is performed reliably and repeatedly by well-trained airline pilots and their aircraft in the NAS today. Accordingly, UAS operators performing commercial operations should be required to meet all the certification and equivalent safety requirements of a commercial operator. And the pilots flying the aircraft must meet equivalent training, qualification, and licensing requirements of pilots of manned aircraft in the same civil regulated airspace.

UAS Design Standard Challenges

Unmanned operations are envisioned to afford possibilities and convenience that offer the attraction of a flying technology beyond the conventions and constraints of modern aviation. The reality is quite different; new UAS technology currently lack - but must have the standardization of safely integrated and interoperable certified systems, which the FAA requires of commercial operators in the NAS today. Without mature safety standards and without technologies that have certification standards and regulations, safety in the WAS may be significantly and negatively impacted, introducing more vulnerability and risk to commercial airline operations and to an overburdened Air Traffic Control system.

Currently, the UAS technologies, safety, and certifications for an end-to-end solution for WAS integration are immature. Patience and, most importantly, collaboration are needed to diligently examine all of the barriers and successfully develop comprehensive and fully mature solutions prior to widespread operational implementation into the NAS, We simply cannot afford to miss critical steps in technological design standards and safety analyses in an attempt to hastily satisfy a market demand, because the impacts to the safety of the WAS could be profound, far outweighing any benefits.

The introduction of multiple variations of UAS without first completing comprehensive risk analysis, rigorous testing, and robust aircraft and pilot certifications would set back the progress accomplished in aviation safety while simultaneously losing the public's trust of safe air travel. We believe that all aviation stakeholders should examine UAS integration to determine how these platforms may impact their operations.

Technological Challenges Impacting Operations in the NAS

A June 20, 2014, newspaper articlel reported that 47 UAS accidents involving U.S. military and federal agencies' aircraft have occurred since 2001 - a safety record that no commercial business or airline could survive. These federal institutions have the authority to self-certify the airworthiness of their own UAS, which can involve modifying compliance with FAA certification standards to accommodate these agencies' unique mission requirements. This latitude and difference in priorities relative to commercial aviation is likely a contributing factor to the number of UAS accidents.

As such, it is easily understood that without the FAA's and other safety organizations' experience and collective guidance in aviation safety, lesser airworthiness standards and certification procedures will produce greater UAS accident rates. Moreover, these accident rates expose the importance of developing civil standards tailored explicitly to UAS technologies, airworthiness, and related certifications through established civil procedures.

Unlike their manned counterparts, a primary system on a UAS is the communication and control system (C^2). This system is what allows the pilot to remotely control the aircraft. The system transmits and receives command inputs (e.g., flight controls, navigation, aircraft status, and ATC communications) from the ground station via radio frequency to the unmanned aircraft (UA). The criticality of the C^2 system becomes self-evident, as it is the most vital single-system link depended upon for the UAS to successfully and safely operate. Link failure - which is exactly analogous to the pilot of an aircraft suddenly disappearing from the cockpit - may cause a multitude of unintentional, cascading events. The sole dependence on this vital system is a necessary aspect of UAS operations, but its failure is one of the primary causal factors as to why UAS have had many accidents.

The primary C^2 contributing failures are associated with latency issues - that is, the time between transmission and reception of a command to successfully operate the UAS. Unlike the human on-board pilot, whose control input is instantaneous, latency times can be from 3 seconds to as much as 30 seconds, perhaps more. When flying in the NAS, where immediate communication and required actions are expected to provide separation between aircraft, latency could cause more significant problems for ATC and manned aircraft in that airspace.

The varying degrees of UAS C^2 vulnerabilities and failures creates complex safety issues for UAS integration. The C^2 data, voice, and video requirements placed on operating UAS using radio waves or satellite create limitations that currently prevent UAS from performing to the safety level of manned commercial aircraft operations. if a UAS cannot maintain a C^2 link, the normal expectation of a UAS to perform the critical functions of ensuring separation from terrain, obstacles, and other aircraft, as well as collision avoidance responsibilities, will unduly place safety burdens on other NAS users.

Manned aircraft flown by pilots in the NAS today use instrument flight rules (IFR) to take advantage of the benefits of FAA's ATC separation services. However, a pilot's responsibility to see and avoid to remain well clear of other aircraft is a constant responsibility in the pilot's line of work, regardless of who or what else is monitoring the flight. Simply stated, pilots visually scan the airspace, especially when traffic is being reported to them by ATC, to identify the aircraft in question when a traffic alert is initiated or simply when a flight crew is flying into an airport that may not have a control tower to avoid all potential conflicts. The UAS needs to be equipped with the technological ability to maintain well clear of and I "When Drones Fall from the Sky," *Washington Post,* June 20, 2014_k avoid collisions with other operators if it is to truly replicate the actions expected of every aircraft in the NAS.

A robust and safe UAS design should never result in the transference of safety responsibility - such as that for maintaining separation - to other operators and users. Accordingly, one of the most important capabilities yet to be developed for UAS operations is the detect and avoid (DAA) technology that is fully capable of performing two primary functions: staying well clear of other aircraft and, if that fails, retaining the ability to avoid an imminent collision using an active collision avoidance technology. While those capabilities in manned aircraft are accomplished by a combination of pilot skill and electronic means, UAS must solely rely on electronic means. The responsibility to avoid coming hazardously close to other aircraft is a shared responsibility.

In addition to the UAS ability to detect and avoid aircraft, other aircraft in the NAS must likewise be able to "see" any UAS that could pose a collision threat. Realistically, given sizes too small to be seen by the human eye until the aircraft is dangerously close, the ability to be seen must be done by electronic means.

A promising system to enable that capability is aAircraft cCollision aAvoidance sSystem for NextGen, or ACAS X. Currently in the research and development phase, only limited funding exists to develop and implement this groundbreaking technology. Additional funding for ACAS X (for manned aircraft) and ACAS Xu (for UAS) would accelerate this new capability. This new technology will likely play a critical role in the safe integration of UAS platforms into the NAS, as well as ensure harmonization of UAS with NextGen requirements.

GEOGRAPHICALLY LIMITING TECHNOLOGY FOR UAS OPERATIONS

Technology exists to limit the geographical and vertical limits of unmanned aircraft operations, independent of the performance capability of the aircraft itself. Geographic and vertical limiting of UAS should be required for all UAS that are not intended to "mix" with conventional aircraft or in the vicinity of airports and other sensitive areas, regardless of whether the UAS is flown for business or recreation. Until the FAA mandates the use of such technology, the effectiveness of this solution will be somewhat limited.

Unfortunately, a software-based solution such as geographical and vertical fencing will be subject to hackers, or those intent on defying the regulations.

Attempts to defeat such technology must be viewed as a deliberate act intended to create a hazard in the NAS and dealt with accordingly. Intentionally operating any aircraft whether manned or unmanned, in an unsafe manner is not a hazard to be mitigated - it is a deliberately unsafe act that, like intentionally shining a laser at an aircraft, cannot be tolerated and must result in an appropriate civil and/or criminal penalty.

SUMMARIZING ALPA'S VIEW ON UAS DESIGN, CERTIFICATION, AND OPERATIONS

The pressure for rapid integration of UAS into the NAS must not result in incomplete safety analyses or technologies prior to any authorization approvals to operate. The urgency to allow UAS into the NAS with immature technologies and lack of appropriate standards and certifications at this time should not encumber other NAS users with additional safety burdens. Standards and technologies for UAS must be in place to ensure the same high level of safety as is currently present in the NAS before a UAS can be authorized to occupy the same airspace as airlines, or operate in areas where UAS might inadvertently stray into airspace used by commercial flights. It is critical that the decisions being made about UAS airworthiness and operational requirements fully address safety implications and complete interoperability functionalities (e.g., detect and avoid capability) of these aircraft flying in, around, or over the same airspace as manned aircraft and, more importantly, airline aircraft.

A well-trained and experienced pilot is the most important safety component of the airline system. The role of the pilot is a major area of concern within the UAS - and within the piloted-aircraft communities. UAS pilots should not be allowed to operate UAS in any commercial operation using non-licensed or private pilots.

It is impossible for a UAS pilot to react to anything other than an explicitly annunciated malfunction. A pilot on board an aircraft can *see,* feel, smell, and hear many indications of an impending problem and begin to formulate a course of action before even sophistitated sensors and indicators provide positive indications of trouble. This capability is necessarily lost without a pilot on board, so the margin of safety the pilot represents must be replaced by other means. UAS pilots should be trained, qualified, and monitored to meet the equivalent standards of pilots who operate manned aircraft in either private or commercial operations.

ALPA Recommendations Pertaining to UAS Design and Operations

1. A comprehensive, proactive UAS safety program should incorporate technology standards, safety analyses, certifications, and flight standards to ensure that introduction of UAS into the NAS will not degrade the existing NAS target level of safety.
2. Federal aviation regulations that specifically address UAS operators, operations, aircraft, and pilots must be developed. Any UAS-unique or UAS-specific regulations must be comparable and compatible with other existing regulations for other airspace users.
3. UAS are inherently different aircraft from manned aircraft and should be required to be equipped with safety-based technologies designed with both well clear and aActive collision avoidance functionalities at the heart of their system architectures to operate in normal and abnormal modes and conditions to maintain the current level of safety in the NM.
4. Commercially operated UAS should be flown by pilots who hold a commercial license and an instrument rating to ensure the continuity of safety that now exists in the NAS.
5. Regulations containing certification standards, continuing airworthiness standards, and minimum equipment list requirements for UAS that are intended to operate commercially in the NAS must be developed.
6. Any person or persons in direct control of a UAS must be limited to the control of a single aircraft unless operations are conducted in special-use airspace.
7. Congress should work with industry stakeholders to develop an appropriate UAS integration funding mechanism.

NEAR-TERM CALL FOR ACTION: A FOUR-PART SOLUTION

With the anticipated sale of 1 million UAS hi the next 90 days, there is an immediate sense of urgency that must be considered. ALPA believes that a significant step toward the eventual solution to safely integrating UAS into the NAS includes four fundamental elements:

1. *Education:* Anyone who plans to fly UAS must understand the aircraft, the airspace, and the other aircraft that could be encountered while flying,
   In the case of UAS that might be flown for compensation or hire in civil airspace, the pilot must hold a commercial pilot certificate to ensure he or she possesses the appropriate skill and experience to meet safety standards designed to protect the flying public.
   Those flying UAS for recreational purposes must adhere to the FAA guidelines including potential minimum age requirements, keeping the UAS within line of sight, and flying at heights under 500 feet.
   ALFA urges Congress to provide definitive authority and remove any ambiguity about the extent to which the FM has the authority to regulate sUAS operated for recreation and hobby. However, in the absence of congressional clarification, we believe FAA may be able to utilize its authority to ensure the safety of the national airspace by regulating all aircraft operations including recreationally flown UAS. ALPA stands ready to assist the agency in the swift development of these regulations and help achieve our shared goal of ensuring the safety of air transportation.
   Based on what the FAA has documented to date, the ongoing educational efforts under way by the FAA and recreational UAS segment are woefully inadequate. ALPA remains willing and able to assist.
2. *Registration;* Gathering basic information about the identity of the individual purchasing the VAS not only allows law enforcement authorities to identify the owner if the UAS were to encounter a problem, but it helps make clear the serious nature of operating a UAS in the U.S. WAS and the responsibility to safeguard public safety,

3. *Technology:* If UAS, either intentionally or unintentionally, are operated in airspace that airliners use, airline pilots need to be able to see them on cockpit displays, controllers need the ability to see them on their radar scopes, and UAS must he equipped with active technologies that ensure that the UAS is capable of avoiding collision with manned aircraft. In these types of operations, technology must enable the pilots to control and interact with them in the same manner as if the pilot were on board, if a UAS is restricted by regulations from operating in a particular geographic area and/or altitude, it must have technology that cannot be overridden that limits the geographic areas and altitude in which it can operate. This may include permanent locations such as the White House and all public airports, as well as temporary restrictions such as for wildfires or natural disaster areas.
4. *Penalties and enforcement:* UAS pilots must be properly trained and understand the consequences of possible malfunctions. Anyone flying a UAS that is a hazard to other aircraft in the airspace, especially those who choose to do so recklessly near airports, must be identified and appropriately prosecuted. We support the criminalizing of intentionally unsafe operation of UAS and penalties for unintentional unsafe VAS operations. If additional funding is needed for this purpose, Congress should provide the resources needed without delay.

CONCLUSION

American aviation technology is experiencing its own "space race" akin to the 1960s. With phenomenal growth in aviation science and technological advancements in this modern digital age, the results are testimony to the advanced applications underpinning NextGen and associated programs. These technologies are designed at their core architectures to be safe, reliable, and repeatable to provide the efficiencies required to maintain the target level of safety as aviation transportation continues to grow.

The target level of safety for commercial air travel in the NAS should be proactively, not reactively, protected. We are fully aware that there is a strong desire by VAS proponents, and those who wish to become operators, to begin flying in the NAS as quickly as possible. Clearly, there are commercial, social, business, and international competitive advantages to a strong VAS industry.

However, government and industry must take a longer view of this present state of technology to ensure that robust safety systems, in tandem with FAA-certified redundant systems of UAS, are developed that completely integrate with commercial airline operations and, above all, do so safely. An Imprudent rush to create and implement minimum standards will not only harm safety, but potentially produce a setback for the future expansion of UAS operations for years to come.

Data show that UAS sales are skyrocketing while hazardous UAS encounters are also rapidly escalating. The need for immediate action is clear; but without comprehensive certification and operational rules and policies, the challenges we face will only continue to multiply.

On behalf of the more than 52,000 pilots whose top priority is safe transportation, we thank the committee for the opportunity to testify on this important subject and look forward to working together to ensure the safety of our air transportation system.

In: Unmanned Aerial Systems
Editor: Wilfred Sparks
ISBN: 978-1-63484-652-3

Chapter 10

STATEMENT OF RICHARD HANSON, DIRECTOR OF GOVERNMENT AND REGULATORY AFFAIRS, ACADEMY OF MODEL AERONAUTICS. HEARING ON "ENSURING AVIATION SAFETY IN THE ERA OF UNMANNED AIRCRAFT SYSTEMS"*

Chairman LoBiondo, Ranking Member Larsen and members of the committee, thank you very much for the opportunity to participate in today's hearing on aviation safety. I'm speaking on behalf of the Academy of Model Aeronautics (AMA), the world's largest community-based organization[1] representing those who fly small unmanned aircraft systems (sUAS), or model aircraft as we call them, for recreational and educational purposes. For nearly 80 years, we have been dedicated to promoting and preserving the hobby of flying model aircraft.[2]

We advocate for and support our more than 180,000 members and 2,400 chartered clubs across the country. The AMA provides leadership, organization, competition, protection, representation, education and scientific/technical development to the model aircraft community. The truth is that hobbyists have flown model aircraft since before manned aviation began.

* This is an edited, reformatted and augmented version of a statement presented October 7, 2015 before the House Committee on Transportation and Infrastructure, Subcommittee on Aviation.

Hobbyists have over 100 years of experience flying model aircraft, and the hobbyist community has established a legacy of safe and responsible flying.

As you're well aware, we are currently seeing unprecedented growth in the development and expanse of unmanned aircraft. According to the Consumer Electronics Association, drone sales in the U.S. will reach 700,000 units in 2015 – a 63% increase from the previous year.[3] The growth in this technology and the emergence of the supporting industry is exciting and extremely beneficial to our economy and our society as a whole. This technology brings with it beneficial applications in areas of exploration, preservation, conservation and humanitarian endeavors. Unmanned aircraft can also be extremely beneficial in the event of manmade and natural emergencies. This technology has the potential to improve our quality of life, to save lives, and to prevent injury in hazardous and perilous situations.

As with any emerging technology, the advent of UAS brings with it policy challenges. Challenges in terms of balancing safety, sustained industry growth, and public benefits. While technology solutions are important tools, enhanced educational efforts through community-based organizations, such as the AMA, as well as stricter enforcement of existing regulations and establishing the regulatory framework around commercial use of this technology, will make the greatest strides towards increased safety.

It's been suggested that technology itself can be the solution to the concerns relating to UAS. Although technology can certainly be part of the solution, relying solely on technology in a snapshot of time could ultimately have a stifling effect on future development and may prove to be ineffectual as new technologies emerge.

One of aviation's primary safety tenants is the final authority and responsibility of the operator to assure the safe operation of his or her aircraft. Relying solely on technology to ensure the safety of flight effectively shifts responsibility and accountability from the operator to the manufacturer and/or third party software developers. AMA strongly suggests that technological solutions be geared toward enhancing the operator's situational awareness and providing the information necessary to make informed and prudent operating decisions. Technology that overrides or impedes the operator's discretionary actions may have the effect of inhibiting legitimate operations, could inadvertently cause an unsafe operating condition and may ultimately hinder the operator's ability to manage an emergent situation.

To the industry's credit, many of the leading companies that manufacture unmanned aircraft for both recreational and commercial purposes have developed and implemented technology to address some of these issues and safeguard against the risks and threats we are all concerned about. These include default altitude limitations and GPS-based information[4] (often called geo-fencing) to help prevent inadvertent flight by operators who may not be familiar with airspace restrictions. Industry innovations include an automatic return-to-home feature that engages when the command signal is lost or when the battery runs low. As circumstances and issues evolve, the industry is rapidly developing additional technologies to address safety and operational concerns, not only because it is in their interest but also because their customers are demanding these features.

While the use of this technology is gaining notoriety and has generated questions regarding safety and appropriate use of UAS, the traditional model aircraft community continues to operate recreational UAS in safe and responsible manner. AMA's members in particular have an unparalleled safety record. The Academy first began developing community-based safety guidelines[5] when it was founded in 1936, which actually predates the FAA by 22 years. These educational tools are critical for our members to learn and stay up to date about the safety aspects of operating their model aircraft. AMA's educational programs teach recreational users not to interfere with and to remain well clear of manned aircraft, to know where and when it's inappropriate to fly their model aircraft, and to notify airport authorities and air traffic control when flying in close proximity to an airport.

All of our members must fly their model aircraft in accordance with AMA's National Model Aircraft Safety Code[6] and affirm their voluntary compliance with the Academy's safety guidelines and those of their local flying site as a condition of membership. AMA's community-based safety program has been recognized by Congress as well as state legislatures as a safe and effective means of managing the operation of model aircraft and the activities of the aeromodeling community. The AMA has an infrastructure of organizational leadership and procedures that regularly reviews techniques, technologies, ideas and guidelines that would affect the safety of model aircraft use. This process keeps the safety code and modeler practices in sync with current practices and a changing regulatory environment.

As a result, today AMA's safety guidelines continue to evolve to accommodate new technologies, new modeling disciplines, and a diverse community of model aircraft enthusiasts from six to 96 years old who safely enjoy this fun, educational hobby. AMA has always believed that the best, and perhaps the only, way to successfully manage the recreational community is through a community-based set of safety guidelines. This system has worked for decades, is working today, and should continue for many generations to come.

Protecting model aircraft enthusiasts is about much more than a hobby – model aircraft is also an effective teaching and career development tool. During the Apollo program of the 1960s, the average NASA engineer was in his or her late 20s. Today's average aeronautical engineer is in his or her mid 50s. Sadly, numerous technological innovations of today are developed outside the United States. The need is clear: America must foster a new generation of engineers, scientists, technologists, and skilled individuals who are relevant in the technology-based workplace.

And model aviation has long served as a stepping stone for these careers. The majority of America's civilian, military, and commercial aviators, as well as aviation professionals, report being influenced by model aviation early in life. Neil Armstrong, aviation pioneer Burt Rutan, and astronaut Robert "Hoot" Gibson—all lifelong AMA members—are just a few examples. Even "Miracle on the Hudson" Flight 1549 pilot Chesley "Sully" Sullenberger built model airplanes in his youth.

In support of the effort to foster a new generation of engineers, AMA as a national organization and our local chartered clubs across the country are involved in and support many science, technology, engineering and mathematics (STEM) projects. There's no doubt that model aviation is fertile ground for the innovations that will one day be tomorrow's technology. Consequently, we must be careful to protect the fun and educational hobby of model aeronautics.

Unfortunately, protecting the hobby of model aircraft is no easy task in today's regulatory and media environment. Drones, as UAS are commonly named, have become increasingly popular and readily available to the general public. Like many new technologies, this proliferation has created concern about safety.

Headlines from the last several months have reflected this increased focus on drones. Stories portray drones "clogging U.S. airspace," "snarling air traffic," "giving the FAA fits," and "penetrating some of the most guarded airspace in the country." As the largest national organization representing long-time hobbyists, we have been very concerned about these reports. Safety is the cornerstone of our organization and our members have been flying safely for many years.

In order to better understand the current state of drones in the United States, and what role we could play to help advance safe and responsible flying, AMA closely analyzed each of the more than 700 records of "drone sightings" released by the FAA this summer. Our analysis of the FAA's drone data reveals a very complex picture of drone activity in the U.S.[7]

Without a doubt, the FAA's data includes some records of "near misses" that represent actual safety concerns, and more needs to be done to address those. At the same time, our analysis shows that the number of "close calls" and "near misses" is substantially lower than the media headlines would suggest. While a single near miss involving a UAS is cause for concern, exaggerating and mischaracterizing the data makes it impossible to assess the magnitude of the situation and appropriately address the risks involved.

A closer look at the FAA's data reveals a hodgepodge of reports. First of all, not every sighting or drone report found in the data was truly a "close call." Many were just that – sightings. AMA's analysis of the data shows that only a fraction of the records were legitimately reported as "near misses" or "near midair collisions." In fact, only about 3.5 percent of the reports were identified with explicit notations as a "near miss" or a "near mid-air collision." According to the data, the number of encounters that pilots considered near misses was in the dozens, not the hundreds.

Some of the most serious incidents in the FAA's dataset – including two actual crashes – involve government-authorized military drones, not civilian drones. The records also include several reports of authorized or unauthorized public entities and commercial operators flying. Given the widespread interest in commercial applications, unidentified operators cannot and should not be presumed to be "hobbyists."

Some additional sightings appear to involve people flying unmanned aircraft systems (UAS) responsibly and within current recreational guidelines. And even more interesting, AMA's analysis found that many things in the air – from balloons and birds to model rockets and mini blimps – are mistaken for, or reported as, drone sightings even when they are not. Clearly, more analysis of the data is needed.

Before we move forward to address safety concerns, we must first do more to understand what is actually occurring in the nation's airspace. We believe the FAA's drone data could help guide future policy conversations on this issue, and help all stakeholders identify solutions to mitigate true safety risks. But this is only possible if the agency takes the time to analyze and accurately categorize the data.

AMA has worked closely with the FAA for many years, and we continue to consider the agency a partner in promoting model aircraft and consumer drone safety. While the FAA needs to do a better and more accurate job analyzing the problem, there are unquestionably some safety risks that deserve attention. And AMA has several recommendations for what we can do right now to ensure the continued safety of our nation's airspace.

One of the most immediate and helpful things the FAA can do to increase safety is to finalize and implement its small UAS rules. These rules have been long in coming and the FAA recently missed its congressionally-mandated deadline of September 30, 2015 to integrate unmanned aircraft systems into the national airspace. As they are currently written, the proposed rules will enhance safety by requiring everyone who wants to fly to either follow the safety programming of a community-based organization (CBO) like AMA or follow new FAA rules for commercial operators. Once implemented, the new rule puts in place some level of oversight and education for all sUAS operators.

At the same time, the FAA should step up enforcement of existing regulations and work more closely with local law enforcement to pursue bad actors. There are existing federal prohibitions against interference with manned aircraft operations, and regulations that address careless and reckless flight. The FAA has the authority to fine operators up to $25,000 for violations that rise to the level of endangering persons and property.[8] In many jurisdictions, there are also criminal laws that address endangerment, public disturbances, and malicious activity under which irresponsible drone operators could be prosecuted. However, despite the FAA and local authorities having the ability to levy hefty penalties, very few enforcement actions have taken place.

AMA supports the FAA taking a more aggressive approach to enforcement against truly careless and reckless behavior. In AMA's analysis of the FAA's drone data, we also found that more than 140 drone sightings either weren't referred to local law enforcement at all or it is not known whether local law enforcement was notified.

While not every report is a serious safety risk, or even someone behaving irresponsibly, the only way to identify the truly careless and reckless UAS operators and to learn the facts about what happened is better communication and coordination with local law enforcement. Stricter enforcement and increased coordination with local authorities will go a long way to deter irresponsible behavior, and it's a step the FAA can take today.

The most meaningful step we can all take to address safety concerns is to increase education. AMA members, who have been flying for many years, know where and where not to fly, and do so safely and responsibly. Unfortunately, many newcomers to the technology may not be aware of the safety considerations involved in operating their unmanned aircraft or "drone".

That's why AMA has expanded its educational efforts beyond its members to include newcomers to the hobby and those who might be purchasing a drone for the first time.

In December 2014, AMA launched "Know Before You Fly" along with the Association for Unmanned Vehicle Systems International (AUVSI), the Small UAV Coalition and in partnership with the FAA.[9] This education-focused campaign works to put important safety information and flying tips in the hands of new flyers. By working with manufacturers and distributors to include safety brochures in product packaging, at the point of sale and online, we are helping to ensure everyone has access to basic safety information and operating guidelines. To date, six manufacturers and distributors – Castle, DJI, Hobbico, Horizon, Yuneec and UAV Experts – have agreed to include KBYF brochures with their products and more are joining each month. Many of the long-standing aviation association and industry advocates in both the manned and unmanned communities such as the Air Line Pilots Association (ALPA), which is also testifying alongside AMA today, Aircraft Owners and Pilots Association (AOPA), the National Business Aviation Association (NBAA), The Experimental Aircraft Association (EAA) and the Airborne Law Enforcement Association (ALEA) have signed on to the Know Before You Fly campaign and have agreed to help distribute the KBYF safety message.

AMA has also shared materials from the FAA, U.S. Bureau of Land Management and U.S. Forest Service – including the National Interagency Fire Center's public service announcement and the U.S. Forest Service's "If you fly, we can't" flyer – with our members who fly UAS for recreational and educational purposes. While our members already know not to fly in and around wildfires, we've asked them to help spread the message more broadly to non-AMA members who might not be as familiar with safety best practices. AMA also coordinated with the modeling industry to distribute these materials through the network of online distributors and the prominent display of safety posters at brick and mortar retail locations in the affected states.

These efforts are important steps toward increasing education for UAS flyers, both new and old, but it is not the only step we can take. The FAA must quickly finish the proposed UAS rules, more strictly enforce existing regulations, and increase coordination with local law enforcement. This will address many of the safety concerns we all have with drones.

AMA is fully committed to working with all stakeholders on these solutions and to ensure the safety of our airspace for all manned and unmanned aircraft. Thank you again for the opportunity to speak today. I look forward to answering any questions the subcommittee might have.

ADDENDUM
A – DJI SAFETY FEATURES
B – "A CLOSER LOOK AT THE FAA'S DRONE DATA"

Safety Features Implemented in DJI Consumer Drones

- Live map showing drone location, orientation, distance, altitude
- Visual/audible low-battery warning with reserve power
- Altitude limitation (398 foot default)
- First consumer drone with geofencing – provides location-based warnings and, in some cases, restricts flight
- Automatic return-to-home in event of radio signal loss
- Automatic return-to-home in event of low battery
- Automatic user-initiated return to home in the event of any contingency
- Auto-land feature for all of the above return-to-home functions
- Talking voice feature for audible cues when pilot is looking up
- Prediction of battery power needed to return home (avoids "point of no return" problem)
- Motors do not spin until deliberate arming sequence is input by user
- Intelligent battery system with charge history and temperature monitoring
- Built-in flight simulator for practice
- GPS hold feature keeps drone in place even with pilot "hands off" or in wind
- GPS using two different systems for redundancy and reliability
- Vision positioning system for position-hold indoors or at low outdoor altitudes
- Speed limitation (Phantom/Inspire)
- Weak radio signal warning before signal loss
- Live HD view for collision hazard detection and precise positioning
- Auto-tightening or locking propellers eliminates loose ones flying off
- Propeller guards for Phantom series
- Lightweight materials and lowest-in-class drone weight to reduce risk

- Collision avoidance technology for the Matrice 100 (first "retail" drone to offer this)

A Closer Look at the FAA's Drone Data

From military crashes to a UFO sighting, AMA analysis reveals a more complex picture of drone activity in the United States – and only a small fraction were legitimately reported as "close calls"

Executive Summary

Headlines from the past few weeks are enough to make you rethink your summer vacation. "FAA records detail hundreds of close calls between airplanes and drones," proclaimed *The Washington Post*. "Leaked FAA report shows almost 700 close calls between drones and planes," wrote *The Christian Science Monitor*. Stories portray drones "clogging U.S. airspace," "snarling air traffic," "giving the FAA fits," and "penetrating some of the most guarded airspace in the country."

As a nationwide community-based organization of more than 180,000 model aviation enthusiasts, the Academy of Model Aeronautics (AMA) is deeply concerned about these reports. Safety is the cornerstone of our organization; our members have been flying model aircraft safely for nearly eight decades.

In order to better understand what's occurring, and what role AMA could play to help advance safe flying, our organization closely analyzed each of the more than 700 records of "drone sightings" recently released by the FAA.

Without a doubt, some drones are flying too close to manned aircraft, airports, wildfires, critical infrastructure and in restricted airspace. AMA is concerned about these reports and helped create the 'Know Before You Fly' (KBYF) campaign in 2014 to educate newcomers to drone technology about where they should and shouldn't fly. AMA and its partners continue to work with manufacturers and distributors to include safety brochures in product packaging and/or at the point of sale. To date, six manufacturers and distributors – Castle, DJI, Hobbico, Horizon, Yuneec and UAV Experts – have agreed to include KBYF brochures with their products and even more supporters in the manned and unmanned aviation communities are joining each month. Hobby People, a brick and mortar retailer, is displaying the KBYF materials at the point of sale in all 18 of their stores. DJI, which manufacturers the popular Phantom quadcopter, has also asked its sales dealers to distribute KBYF brochures with drone equipment sales, and has implemented altitude limitations and GPS-based warnings and limitations into its products.

Beyond education, AMA has encouraged the FAA to more aggressively enforce existing rules against careless and reckless behavior, as well as violations of restricted airspace. The FAA currently has the authority to assess civil penalties of up to $25,000 against careless and reckless operators. Hefty fines could help deter bad behavior, yet very few fines have been levied to date. While AMA's members are responsible and know where they should and should not fly, all users of the airspace have a responsibility to ensure safety, and AMA is committed to doing its part.

At the same time, AMA's analysis of the FAA data shows that the number of "close calls" and "near misses" is substantially lower than the headlines would suggest. So what's in the data? A closer look reveals a hodgepodge of reports. Some of the key takeaways:

- Not every sighting or report was a "close call." Many were just that – sightings. Only a small fraction was legitimately reported as "near misses" or "near mid-air collisions."
- Some of the most serious incidents in the FAA data – including two actual crashes – involve government-authorized military drones, not civilian drones.
- It's not just uninformed consumers causing problems; the records include several reports of authorized or unauthorized public entities and commercial operators flying. Given the widespread interest in commercial applications, unidentified operators cannot be presumed to be "hobbyists."
- Some sightings appear to involve people flying responsibly and within the FAA's current recreational guidelines.
- Many things in the air – from balloons and birds to model rockets and mini blimps – are mistaken for, or reported as, drone sightings even when they are not. One pilot in Minnesota even reported seeing something that "resembled a dog."
- A number of sightings have occurred over or around stadium events, wildfires, power plants and other critical infrastructure. These raise different concerns from pilot sightings.

- Despite the FAA's stated desire to find and punish rogue operators, in almost 20% of reports – 142 reports, to be exact – local law enforcement either wasn't notified or it was unknown whether local law enforcement was notified.

The following analysis delves into each of these findings in greater detail. Meanwhile, the conclusion contains recommendations for the FAA going forward, both in terms of how this data should be released in the future as well as what more can be done to address instances of irresponsible operations.

The Majority of These Reports Are Sightings. Only a Few Dozen Are Explicitly Reported as "Near Misses" or "Near Mid-Air Collisions"

There's a big difference between a sighting and a near miss. Seeing a drone doesn't necessarily mean there was a "close call" or that there was any safety risk. We analyzed each of the FAA's 764 records in an attempt to discern the true "near misses" from what may be more appropriately be called "sightings." We tallied up the number of narrative reports that explicitly called an event a "near miss," "near collision" or NMAC (near midair collision). We also looked at when a pilot took evasive action to avoid a drone and when no evasive action was taken, or evasive action was unknown. Here's what we found:

- Despite the perception that all of the 764 reports are "close calls," we only identified 27 reports – or 3.5% – that had some explicit notation of NMAC, "near miss" or "near collision."
- Several reports explicitly indicate the opposite – that a pilot did not consider the event a near miss. These notations include:
  - "PILOT ISN'T REPORTING A NEAR MID-AIR"
  - "THE PILOT DID NOT CONSIDER IT AS A NMAC"

Since so few reports explicitly indicate whether a near miss did or did not occur, we drilled down further and examined how many pilots reported taking evasive action in response to a drone. The numbers are also surprisingly few:

- In 51% of the records (392 or 764), there is an explicit notation of "No Evasive action taken," "Evasive action unknown" or a variation thereof.
- Only 1.3% of the records (10 of 764) explicitly note that a pilot took evasive action in response to a drone.
- The rest of the records do not address evasive action, conflicts or near misses at all, either because of an omission in the data or because the remaining reports do not relate to pilot sightings.

Evasive action taken?	Number of records	Percent of total reports
No evasive action taken or evasive action unknown	392	51.3%
No reference at all or not applicable	362	47.4%
Evasive action taken	10	1.3%

AMA's Recommendation(s): The FAA needs to better analyze and categorize pilot reports to indicate which present serious safety risks (near mid-air collisions) and which could be more appropriately classified as sightings. There are several issues with the FAA's handling of this most recent announcement of drone sightings.

First, the FAA fueled inaccurate and sensational headlines claiming that all drone reports involve "close calls," which even a causal review of the FAA's data shows is inaccurate.

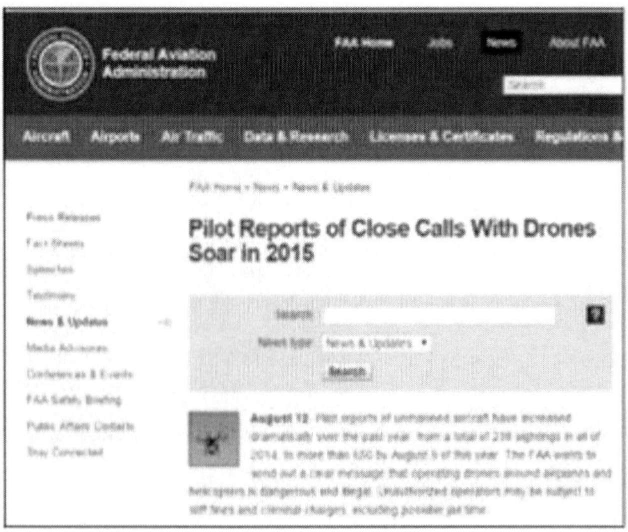

This happened first and foremost because the agency's own press release from August 12, 2015, bears the headline, "Pilot Reports of Close Calls With Drones Soar in 2015" and the FAA cited more than 650 reports.[1] When the actual reporting data was released on August 21, the FAA's characterization changed dramatically to describe these same reports as merely "possible encounters." During that intervening two-week period, public opinion was influenced by media coverage that relied upon, and cited, the FAA's description of these 650 possible encounters as "close calls." These media reports have, in turn, been cited by lawmakers in Washington and in various states as a reason to call for new legal restrictions.

More recently, the FAA admitted in an August 30 news story that there is no regulatory definition for what constitutes a "close call." The agency said the phrase was "simply part of a news headline,"[2] Implying that the agency may have intentionally used sensational language to spur media interest in its data. The general public, the media, our nation's lawmakers, and all stakeholders in drone technology deserve better from the government agency responsible for the safety of the nation's skies and for reporting objectively and accurately on public safety issues.

FAA's characterization of its data has led to sensationalized and inaccurate reporting, including in The Washington Post.

Most importantly, the FAA's lack of analysis makes it more difficult for the broader community, whether recreational or commercial, to easily identify the most serious safety risks and work toward advancing solutions. There is a flattening of perception that every report is just as important as the next, which clearly isn't the case. Some reports in the data – where near misses occurred or when pilots took evasive action – deserve more attention than others.

Meanwhile, the release of preliminary, raw data without any analysis, context or investigative findings only tells part of the story. Presumably there are investigative findings and other information on at least some of these sightings that would shed light greater on their accuracy or legitimacy. The FAA should be more transparent with this information.

MILITARY DRONES, NOT CIVILIAN DRONES, ARE INVOLVED IN SOME OF THE MOST SERIOUS INCIDENTS, INCLUDING ALL ACTUAL CRASHES

While civilian drones have been the focus of news reports, more than a dozen events in the FAA data involve military drones, or are very likely FAA-approved military drone operations. What's more, the incidents involving military drones are among the most serious in the data, including two actual crashes.

- On July 24, 2015, Lockheed Martin lost control of a small military drone in Owego, NY, prompting the operator to alert a nearby control tower to "watch out."
- On May 26, 2015, a Swedish Air Force C-130 climbed to avoid a military MQ-9 Reaper drone near Victorville, CA.
- On March 25, 2015, an MQ-1C military drone crashed near Wilsona Gardens, CA. The MQ-1C, manufactured by General Atomics, is reportedly an *upgrade to the popular Predator drone.*[3]
- In what may be the same event as above, on March 25, 2015, a citizen in Lancaster, CA reported that a "25 foot drone crashed" in a residential neighborhood. A *photo* was posted to the website of a local radio station.[4]
- On March 3, 2015, controllers at MacDill Air Force Base in Tampa, FL reported that a drone weighing "less than 300 pounds" crashed by the perimeter fence of the base.

While the incidents detailed above are among the most serious, there are several other reports involving military drones, or possibly military drones.

- A Cessna pilot in Van Nuys, CA reported a "huge, silver, military style UAS" which passed a few hundred feet overhead.
- A U.S. Army aircraft operating near Washington, D.C. reported seeing a drone that is "dark gray and smaller than a Predator."
- A VFR pilot in Anchorage, AK entered "hot" restricted airspace where a military drone was operating.

The number of reports involving military drones, or possible military drones, calls into question how many other reports in the FAA data may have been sightings of military aircraft. This is especially true of sightings at higher altitudes, including Class A airspace, which is the airspace between 18,000 feet and 60,000 feet above sea level. Reaching Class A airspace is generally beyond the capabilities of model aircraft and consumer-grade drones. Indeed, the most popular consumer drone sold today, the DJI Phantom, has a built-in maximum altitude function.

The FAA data includes reports of drone sightings at the following unusually high altitudes:

- 24,000 feet – July 24, 2015, near Colorado Springs, CO
- 51,000 feet – July 24, 2015 near Washington, D.C.
- 20,000 feet – March 5, 2015 near Los Angeles, CA
- 25,000 feet – February 9, 2015 near Teterboro, NJ
- 25,500 feet – February 9, 2015 near West Palm Beach, FL
- 19,000 feet – November 24, 2014 near Beaumont, TX

To put these reports into context, many small *manned* aircraft cannot even reach these heights. For example, the service ceiling – the maximum achievable altitude – of a single-engine Cessna Skyhawk (C172) is 14,000 feet.[5] Likewise, the maximum operating altitude of a manned Robinson R22 helicopter is 14,000 feet.[6]

AMA's Recommendation: The FAA should separate out reports that involve military or government agency-operated drones, or that likely involve such drones, from future drone sighting data. This could either mean excluding these military reports entirely, or explicitly indicating when the report relates to military use, and not civilian use, of a drone. These incidents should not be included in reported numbers of civilian drone sightings.

There Are Public Entities and Commercial Operators in the FAA's Reports Too

The FAA currently prohibits commercial drone use unless the operator has received a Section 333 exemption to fly. Likewise, public entities wanting to fly drones – including publicly funded universities, law enforcement agencies, fire departments and other government agencies – can do so by applying for a Certificate of Waiver or Authorization (COA) from the FAA. Despite the perception that all irresponsible operators are uninformed consumers or recreational users, the FAA data reveal several instances of public entities and commercial operations – some or all of which may have been unauthorized.

- In perhaps the most surprising example, on August 18, 2015, the Los Angeles Police Department notified the Inglewood Police Department that a drone the latter agency was flying over a crime scene "needed to come down." The drone was flying two miles from the approach end of a runway at LAX.
- On August 5, 2015, the Coast Guard in Mobile, AL reported a drone being used to film a commercial for a car dealership, in close proximity to the Mobile airport. The report notes that the company involved "does not appear to have a section 333 exemption."
- On August 2, 2015, authorities in Pomona, CA observed a drone flying over a concert, 1/4 mile from a nearby airport. The report states the drone was operated by the event host, Live Nation.
- On July 7, 2015, a pilot in Redding, CA reported a drone 1/4 mile from the runway of a local airport. Local police verified that the drone was being operated by a construction company, which was advised to cease operations.
- On April 28, 2015, a local photography and marketing business in Winchester, VA was reportedly hired to capture aerial photos of the Shenandoah Apple Blossom Festival. The same company may have also used a drone to produce a public service announcement for the Winchester Public Schools. The company does not appear to have FAA approval to fly. Meanwhile, the superintendent of the

Winchester Public Schools said the business owner, "does this all the time."

Whether innocent or intentional, there are some public entities and commercial operators in the FAA's data. Meanwhile, as with the military reports, the presence of several clearly identifiable public entities and commercial operators in the reports raises questions about how many other drone sightings may be authorized or unauthorized public entities and commercial operators. All of the sightings like these have been inaccurately and unfairly characterized by the FAA as "hobbyist" operations, when clearly not all are.

AMA Recommendation(s): As with military reports, the FAA should clearly separate out and explicitly indicate reports that are public entities and commercial operators, or very likely public entities and commercial operators. In the absence of any operator identification, and given the heightened interest in commercial applications, it is inappropriate to assume, or to report, that a drone was operated by a "hobbyist."

Meanwhile, the FAA needs to finalize its small UAS (drone) rules without any further delays. Congress directed the FAA to put these rules in place by September 30, 2015, a deadline the agency is likely to miss. The faster these rules are finalized, the faster public entities and commercial operators will have a regulatory framework with clear requirements in which to operate. Rules won't just help public entities and commercial operators; they will enhance safety across the board by requiring everyone to follow either the new rules for commercial operators or the safety programming of a community-based organization (CBO), like the AMA.

Some Drone Flyers Appear to Be Following Proper Recreational Guidelines, Yet They Are Reported Anyway

One of the more surprising findings in the FAA data is the discovery that, in at least half a dozen instances, some drone flyers may be following FAA-endorsed recreational guidelines. Yet, these users are reported and included in the FAA data set because of a citizen or pilot who is unfamiliar with the proper guidelines.

The FAA has indicated that recreational flyers should stay below 400 feet, stay well clear of manned aircraft, fly their drones or model aircraft within visual line of sight at all times and not fly within five miles of an airport without prior notification to the airport operator or air traffic control tower (when an air traffic facility is located at the airport). Below are examples of reports that appear to meet these guidelines:

- On July 24, 2015, a "low flying" drone was reported five miles south of the Leigh Valley International Airport in Allentown, PA. No specific altitude was reported.
- On June 10, 2015, a helicopter flying "below 500 feet" reported seeing a drone five miles away from Fort Lauderdale-Hollywood International Airport.
- On April 28, 2015, a 9-1-1 caller reported a drone 13 miles from Chicago Midway International Airport. The caller stated the drone was "not high."
- On March 28, 2015, a resident of Sioux Falls, SD reported a drone flying "around the area...at 400 feet."
- On February 14, 2015, a drone was reported "below 400 feet" in the vicinity of downtown Anchorage, AK.
- That same day, a helicopter reported a drone 6.5 miles away from Miami International Airport at an altitude of "400-500 feet."
- Also in Miami, on December 14, 2014, a helicopter pilot reported a drone at 150 feet, 5 miles from Miami International Airport.

These are just some of the reports that may actually be cases of safe and responsible operators who are flying within the FAA's current guidelines. There may be many other operators following the guidelines in the data set, but it's difficult to discern because of the two different ways altitudes are reported.

When pilots report drone sightings, they typically report the MSL (mean sea level altitude) of their aircraft along with the sighting. That's because altimeters in manned aircraft measure altitude above sea level.

Drone enthusiasts on the ground are advised to stay under 400 feet AGL (above ground level).

At sea level, 400 feet MSL equals 400 feet AGL. As a practical matter, the ground elevation of most places in the country is higher than sea level, especially in western states. This means that, to get a true picture of how high a drone may be flying above its location, one needs to subtract the ground elevation from an altitude report given in mean sea level.

Consider the following hypothetical example. A drone enthusiast flying six miles away from Denver International Airport (DEN) believes he is following all of the appropriate recreational guidelines, including staying below 400 feet. Meanwhile, a pilot landing at DEN reports a drone off in the distance at 5,700 feet. Is this operator doing something wrong?

No. The pilot is reporting his altitude above sea level, 5,700 feet. The field elevation of the Denver airport is 5,327 feet. That means that the drone is actually flying at 373 feet above ground – under 400 feet. In this instance, there is no "close call" and no danger to the manned aircraft, it is merely a reported sighting of a drone operation that is in compliance with FAA guidelines, which include the instruction that drones should yield right of way to manned aircraft. The direction to yield right of way necessarily means that drones and manned aircraft will be visible to each other's pilots, and that sightings will occur and will be reported.

AMA Recommendation(s): First and foremost, the FAA should exclude from its records any reports where the operator is following, or very likely following, recreational guidelines. There are at least a few of these. By including these reports – and only including preliminary data – the agency is essentially considering all drone operators guilty until proven innocent. The presumption is, if someone sees a drone and makes a report, the drone operator is automatically doing something wrong, which isn't always the case.

These reports also underscore the problems with preliminary data. Some of the reports noted above, as well as others in the data set, may later have been investigated and found to be safe and/or appropriate flights. But the public doesn't know if that's the case because no investigative findings are being released. As stated previously, the FAA needs to be more transparent and release more information, not less.

Finally, in keeping with greater transparency, the FAA needs to clear up the ambiguities in altitude reporting going forward. Specifically, AMA would like to see any altitudes for drone sightings reported in AGL altitude, in addition to the MSL altitudes reported by some pilots. This conversion could be done, or at least approximated, by using the field elevation of the nearest airport. It is also important for pilots to be prompted to include in their reports the estimated altitude of the drone they have observed, not just their own altitude.

Drones Are the New UFOs...Except for One Report of an Actual UFO

In more than a dozen FAA records, pilots either reported seeing "something" other than a drone, or a pilot was unsure whether an object was a drone at all. Such an object is, quite literally, an unidentified flying object ("UFO"). However, it seems the term "drone" has become the new UFO, applying to everything from balloons and birds to model rockets and mini blimps. A few examples:

- On August 12, 2015, a pilot in Salt Lake City, UT reported seeing "an ultra light or drone type aircraft."
- On July 19, 2015, ground personnel at JFK in New York reported a cluster of silver balloons blowing over the airport fence. Controllers in the tower also noted seeing "a silver object."
- Also at JFK, on July 10, 2015, a JetBlue pilot reported seeing "a fast moving gray object."
- On July 5, 2015, an American Airlines pilot near LaGuardia Airport in New York reported seeing a "model rocket" at 2,800 feet.
- On May 31, 2015, a Quantas pilot reported seeing a "mini blimp" at LAX in Los Angeles.
- In one of the more ambiguous reports, on April 5, 2015, an Alaska Airlines pilot near Des Moines, WA "reported seeing something below" his aircraft.
- In what could have been a sighting of a manned aircraft, on February 10, 2015, a Southwest Airlines pilot reported seeing an "unidentified helicopter" while on final approach to LAX in Los Angeles.
- On November 30, 2014, a plane landing at Andrews Air Force Base outside of Washington, D.C. reported seeing a drone that looked like a "large vulture."

There are several more reports that indicate that pilots, at times, aren't sure what they're seeing. "Pilot was not positive that the object he saw was a drone" reads one entry from Sarasota, FL. A pilot in Georgia reported seeing "a red UAS (or balloon)." Similarly, a pilot near San Francisco, CA reported "a UAS or balloon at 10,000 feet."

While any object in the sky seems to be reported as a drone, occasionally a UFO is still a UFO. On July 24, 2015, two aircraft near Washington, D.C.

spotted what they describe as a UFO at 51,000 feet – higher than commercial airliners typically fly.

A UFO Sighting?

"A/c reported a ufo at 150nm north of swl approximate fl510 moving from west to east just above horizon with steady light illumination. Fast moving due east gone within 5 minutes. Ava020 concurred with this report. Report to HQ/DEN."

Even airline pilots who fly very close to these "UFOs" can misperceive what they are. In one report, an airline pilot reported that his aircraft had actually collided with a drone and created sparks. Upon landing, an investigation revealed that his aircraft had actually collided with a bird. With approximately 10 billion birds in the United States, and with many birds presenting an anatomical shape in flight that is similar to the "X" shape of a quadcopter drone, it is quite conceivable that at least some of these reports are actually bird sightings.

It is also apparent that the FAA's reporting system reflected in the summary spreadsheet released to the public on August 21 is not always accurate or complete. For example, the spreadsheet indicates that on January 31, 2015, the pilot of United flight 1087 on approach to Boston's Logan Airport "REPORTED SEEING A UAS 100 FEET ABOVE ACFT ALTITUDE OF 7,000 FEET FOOTBALL SHAPED RED-BLUE IN COLOR." However, the ATC audio recording indicates that the pilot actually reported: "I don't know if it was a balloon or a drone."[7] The FAA's spreadsheet omits any mention of the possibility that this object was a balloon, even though the shape, altitude and pilot's audio recording all point to that as the more likely explanation. This misreporting by the FAA's own system suggests a bias towards reporting flying objects as drones even when the existing evidence indicates otherwise.

AMA Recommendation(s): The FAA shouldn't include reports of flying objects that are clearly not drones – balloons, model rockets, birds, etc. - in its data. The FAA should also separate out and clearly identify reports that are ambiguous – something below the aircraft, dark gray object, etc. – or which express doubt as to whether a flying object is, in fact, a drone. The report should include all facts as reported by the pilot to ATC, FAA or the airline, including any uncertainty about the object that was sighted.

The Data Isn't All About Pilots and Airliners. People Are Flying Over Stadium Events, Wildfires, Power Plants and Other Places That May Raise Different Concerns

While the FAA's press release and subsequent news coverage of the drone data focus on pilot reports and airliners, we have identified at least 26 records of drones flying near stadium events, wildfires, critical infrastructure and in restricted airspace. While these are potentially objectionable if unauthorized, they are certainly not "close calls" when no other aircraft is reported in the area.

Drones over stadiums

According the FAA data, drones have been sighted over the following stadiums and event venues, in some cases more than once:

- University of Alabama Stadium, Tuscaloosa, AL
- High School Football Stadium, Glendale, AZ
- Levi's Stadium, San Jose, CA
- Stanford University Stadium, Stanford, CA
- Nationals Park, Washington, DC
- Marlins Park, Miami, FL
- Fenway Park, Boston, MA
- Bank of America Stadium, Charlotte, NC
- Citi Field, New York, NY
- PNC Park, Pittsburgh, PA
- Seahawks Training Camp, Renton, WA

Source: FAA database on Pilot UAS reports

For stadium events, wildfires and VIP movements, the FAA typically issues temporary flight restrictions (TFRs) that limit access to the airspace for both manned and unmanned aircraft.

For example, the FAA restricts flights over stadiums during Major League Baseball games, National Football League games, NCAA games and motor speedway events. The so-called "stadium TFR" prohibits aircraft operations at or below 3,000 feet AGL within a three nautical mile radius of any stadium with a seating capacity of 30,000 or more. The TFRs go into effect one hour before the scheduled event time and last until one hour after the event concludes.

Specifically, our analysis found:

- At least six reports of drone activity in and around Washington, DC's restricted airspace, including drones incidents at the White House that have been the subject of extensive media coverage.
- In another instance of a security breach, the Coast Guard reported a drone in the vicinity of the President playing golf in West Palm Beach, FL.
- At least 13 sightings involve stadiums or large event venues. However, it is often not noted whether a TFR was in place at the time of the report.
- There are at least four reports of drones flying near power plants in Maine, Wisconsin and New Jersey. It is not clear whether these flights raise an actual safety or security concern.
- There are at least two reports of drones interfering with wildfire operations in California, which have also been the subject of extensive media coverage.

AMA Recommendation(s): While some of these sightings are likely more serious than others, they all speak to the need for greater education so that people aren't flying where they shouldn't. The reports of drones flying around wildfires are among the more serious and AMA has been working closely with the U.S. Forest Service and the Bureau of Land Management (BLM) to distribute educational materials including the U.S. Forest Service's "If You Fly, We Can't posters," to hobby shops out west.

Meanwhile, stadium events are clearly an attraction to drone flyers. At the same time, it is not immediately clear how many of these flights may have violated temporary flight restrictions. Going forward, the FAA should explicitly note whether a TFR was in place during a report that involves a stadium. More broadly, the FAA should identify and separately categorize all events that are, or are very likely, violations of TFRs and other restricted airspace. Despite the concerns these reports might raise, they should not be included in the same category as sightings by pilots.

A Large Number of Reports Aren't Even Referred to Local Law Enforcement

The FAA has said in recent media reports that it is working closely with local law enforcement to identify and investigate reports of unauthorized operations. At the same time, our analysis of the FAA drone data finds that almost 20% of reports – 142 in total – either were not referred to local law enforcement or it is unknown whether a report has been referred to local law enforcement.

142

The number of drone reports not referred to law enforcement, or law enforcement notification is unknown

These reports bear notations such as:

- "NO LEO NOTIFIED"
- "LEO NOT NOTIFIED"
- "UNKN IF LEO NOTIFIED"
- "NO REPORT OF LEOS WERE NOTIFIED"

While the FAA has said that identifying rogue operators is a challenge, in a sizeable number of cases, the agency isn't even attempting to do so. The reasons for this aren't immediately clear. In one instance, a report notes that an incident occurred in Bahaman airspace and therefore law enforcement wasn't notified. However, in the vast majority of these 142 records, no explanation is given. Without such referrals, the possibility of learning the facts concerning these "possible sightings" is virtually zero.

AMA Recommendation(s): The AMA has been calling for more aggressive enforcement of careless and reckless operators and therefore urges the FAA to, as a starting point, report all instances of potentially irresponsible behavior to local law enforcement for further investigation. Our analysis has shown that not all of these reports are equally serious, and some reports may actually be people flying responsibly. That said, it is clear that greater communication and coordination between the FAA and local law enforcement is needed to identify and hold accountable operators who pose a danger to the national airspace system.

CONCLUSION

A close examination of the FAA's drone data reveals a much more complex picture of drone activity in the U.S. There are pilot reports of near misses that represent actual safety concerns, and more needs to be done to address these. But contrary to the FAA's assertion in its press release of August 12, and the widespread media reporting that followed, the narrative descriptions and notations in the 764 reports suggest that the number of actual "close calls" appears to be in the dozens, not the hundreds.

Moreover, the assumption that all drone flyers are "hobbyists" or recreational users is clearly inaccurate. As noted in this analysis, there are several instances of military sightings and mishaps, including two actual crashes. There are also public entities and commercial operators flying with or without authorization. When the operator is not identified, it is not possible to determine the purpose of the operation. Further, some reports may actually be drones operating responsibly pursuant to FAA guidelines, and some reported sightings of drones may not even be drones at all.

There is some useful information in the FAA's dataset – data that could help guide policy conversations about drones and help all stakeholders identify solutions to mitigate true safety risks. But the data is only useful if the FAA takes the time to analyze and accurately characterize it; the same holds true for the media and others.

While AMA works closely with the FAA, and we continue to consider the agency a partner in promoting model aircraft and consumer drone safety, the FAA mishandled the release of its drone data. The agency used misleading language in its press release, released only preliminary reports and did not critically analyze those reports. The agency should not have asserted in the media that there are hundreds of reports of "close calls" with drones when that is not supported by the data, and the agency admits that it has no regulatory definition of what constitutes a "close call."

Moving forward, AMA has two sets of recommendations – one set of recommendations relating to the FAA's handling of its drone data, and another set of recommendations to ensure the continued safety of the U.S. airspace.

Data Recommendations:

1. *Don't just release raw reports or summary numbers; analyze all future drone data.* The AMA believes the FAA needs to conduct an

analysis of all future data releases in order to more accurately portray what is happening. AMA recommends that these reports need to be better categorized going forward to highlight the most serious safety risks so that all stakeholders can work together to address them. Some of the categories or filters AMA recommends include:

- Near misses / near mid air collisions
- Sightings / not reported as a near miss
- Military reports / reports involving military or government agency drones
- Commercial reports / reports involving commercial use
- Potential violations of TFRs or other restricted airspace
- Note which reports are airborne pilot reports vs. ground reports

Additionally, reports relating to balloons, UFOs, model rockets and other flying objects should not be included in the drone data.

In addition to better categorizing the reports, the AMA urges the FAA to provide greater context for these reports, which includes:

- Computing the approximate AGL (above ground level altitude) of any drone reports that are stated in MSL (mean sea level) altitudes.
- Cross-referencing stadium reports with past TFRs (temporary flight restrictions) and explicitly noting whether a TFR was in effect during a stadium-related sighting.

2. *Release not just preliminary reports or numbers, but also investigative findings and any other information.* The problem with preliminary reports is just that – they are preliminary. Once investigated, the reports may turn out to be accurate. However, there's reason to believe that some may be inaccurate. While the FAA only released the preliminary data to the public, The Washington Post reportedly obtained some investigative findings. In one instance, a pilot reported hitting a drone, only later to determine that it was a bird. The Washington Post on August 20, 2015 reports:

> On May 9, the pilot of United Airlines Flight 863 — traveling from San Francisco to Sydney — reported that the Boeing 777 hit a drone at an altitude of 3,000 to 4,000 feet along the California coast.

> "Sparks were observed after contact," according to the FAA report, which said the 777 kept flying because it did not appear to be damaged. A United spokesman said it was later determined that the plane had hit a bird, not a drone.

This incorrect report, from a pilot who was close enough to actually see the object his aircraft collided with, is a good example of why initial reports cannot be relied upon to draw conclusions that might influence policy decisions. The general public, the media, lawmakers, and drone stakeholders deserve greater transparency.

Policy Recommendations

Safety is a top priority for the AMA. Our members have been flying model aircraft safely for nearly eight decades and want to see everyone embrace new drone technology safely and responsibly. While our analysis finds several serious issues with the FAA's data, we are aligned with the FAA when it comes to promoting the safety of the nation's airspace. In light of some of the reports that may constitute careless and reckless behavior, AMA recommends the following:

1. *Refer all reports to local law enforcement.* As noted in our analysis, more than 140 reports either weren't referred to local law enforcement or it is not known whether local law enforcement was notified. While not every report is a serious safety risk, or even someone behaving irresponsibly, the only way to identify the truly careless and reckless operators, and to learn the facts about what happened, is better communication and coordination with local law enforcement.
2. *More aggressively enforce existing rules.* There are existing federal rules against careless and reckless operations, for which the FAA can fine operators up to $25,000. In many jurisdictions, there are also criminal laws under which careless and reckless operators could be prosecuted. Despite the FAA having the authority to levy hefty penalties, very few fines have been assessed to date. AMA supports the FAA taking a more aggressive approach to assessing civil penalties against operators who endanger the safety of the national airspace system.

3. *Finalize the small UAS rules.* One of the most immediate things the FAA can do to increase safety is to finalize its small UAS rules. The draft rules will enhance safety by requiring everyone who wants to fly to either follow the safety programming of a community-based organization (CBO), like AMA, or follow new FAA rules for commercial operators. Once this happens, everyone would need to have some level of oversight and education in order to legally fly. These rules have been delayed several times and, in all likelihood, the FAA will miss its congressionally-mandated deadline of September 30, 2015 to issue final rules.
4. *Continue to educate new users.* Many of these reports underscore the need for more education of newcomers to the technology who aren't traditional model aviation enthusiasts and who aren't trained aviators. AMA is committed to continue advancing safety education under the "Know Before You Fly" campaign along with its campaign partners the Association for Unmanned Vehicle Systems International and the FAA. The FAA itself encourages model aircraft operators: "Do take lessons and learn to fly safely."[8] One of the keys to this commendable suggestion is to recognize that compensation of an instructor's time that is spent helping someone learn how to operate a model aircraft or recreational drone must not be viewed as commercial operation of a UAS requiring a Section 333 exemption and pilot certificate.

End Notes

[1] http://www.modelaircraft.org/uashearing-10-07-2015.aspx.

[2] http://www.modelaviation.com/images/article/AMA75-ONLINESUPPLEMENT.pdf.

[3] https://www.ce.org/News/News-Releases/Press-Releases/2015-Press-Releases/New-Tech-to-Drive-CE-IndustryGrowth-in-2015,-Proj.aspx.

[4] See addendum A – DJI Safety Features.

[5] http://www.modelaircraft.org/uashearing-10-07-2015.aspx.

[6] http://www.modelaircraft.org/files/105.pdf.

[7] http://www.modelaircraft.org/gov/docs/AMAAnalysis-Closer-Look-at-FAA-Drone-Data 091415.pdf See also addendum B – "A Closer Look at the FAA's Drone Data".

[8] https://www.law.cornell.edu/uscode/text/49/46301.

[9] http://knowbeforeyoufly.org/.

End Notes for Addendum

[1] http://www.faa.gov/news/updates/?newsId=83445.
[2] http://www.journalnow.com/news/local/cloudy-faa-data-skews-drone-threat/article 80e10858-3fbb-51ba-827ce5fb55a26534.html.
[3] http://www.ga-asi.com/gray-eagle.
[4] http://www.kfiam640.com/onair/john-and-ken-37487/check-out-this-photo-of-a-13446744/.
[5] http://cessna.txtav.com/single-engine/skyhawk.
[6] http://www.robinsonheli.com/rhc r22 beta ii.html.
[7] http://www.wcvb.com/news/drone-sighted-by-pilot-landing-at-logan/31035050.
[8] https://www.faa.gov/uas/publications/model aircraft operators/.

In: Unmanned Aerial Systems
Editor: Wilfred Sparks
ISBN: 978-1-63484-652-3

Chapter 11

TESTIMONY OF MYKEL KOCHENDERFER, ASSISTANT PROFESSOR OF AERONAUTICS AND ASTRONAUTICS, STANFORD UNIVERSITY. HEARING ON "ENSURING AVIATION SAFETY IN THE ERA OF UNMANNED AIRCRAFT SYSTEMS"*

Chairman LoBiondo, Ranking Member Larsen, and Members of the Subcommittee:

Thank you for the invitation to appear before you to discuss the risks associated with unmanned aircraft, also commonly called drones, in the United States.

I am a professor in the Department of Aeronautics and Astronautics at Stanford University and a third-generation pilot. In this testimony, I am speaking solely for myself.

My research for nearly ten years has involved statistical estimation of risk and the development of technology for enhancing aviation safety. While at MIT Lincoln Laboratory, I helped develop a collection of airspace models, jointly funded by the FAA, DHS, and the Air Force. These models have since been used to estimate collision risk for manned and unmanned aircraft by

* This is an edited, reformatted and augmented version of testimony presented October 7, 2015 before the House Committee on Transportation and Infrastructure, Subcommittee on Aviation.

government, academic, and commercial organizations around the world. My work has also led, in part, to the technology underlying the FAA's next generation collision avoidance system called ACAS X that is currently undergoing international standardization. The FAA is developing a version for unmanned aircraft. My students at Stanford have been supporting this effort and the effort of NASA to build a UAS Traffic Management (or UTM) system.

Personally, I find the rapid acceleration of unmanned aircraft technology to be the most exciting recent development in the field of aeronautics. As we heard in previous panels, this wave of technology has the potential to save lives and create jobs. Few things are as beautiful as flight, and the proliferation of unmanned aircraft has made aviation accessible and inspired a generation of eager university students in a way we have not seen for a long time. It is no surprise that the growing popularity of these vehicles has also raised concern about the safety to other aircraft and the potential for interference with such operations as firefighting and air ambulances. There has been tremendous media coverage of these risks.

I hope to inform the discussion with my thoughts with respect to two questions. First, how do we measure and analyze these risks? Second, what are the best technologies and policies to mitigate these risks?

There is tremendous diversity in unmanned aircraft. We have very small aircraft such as the robot bee developed at Harvard University that weighs about the same as the average honey bee. We also have very large aircraft that are used primarily by the military. I do not believe that this hearing is especially concerned with the risks posed by the extreme ends of this spectrum. Rather, I believe we are concerned with and should be concerned with consumer drones that weigh a couple pounds or more. Drones much lighter than this do not represent true aviation safety hazards due to their physical inability to enter their airspace.

To answer the first question of how to measure and analyze the safety risk of consumer drones, we must have the understanding that risk is determined by both the likelihood and the severity of different hazards.

First, let us consider severity. A sufficiently large drone can cause damage to any part of an aircraft, but one of the most severe hazards is engine ingestion. We all recall the US Airways flight in 2009 that was struck by multiple Canada geese, leading to failure in both engines at takeoff and an emergency landing on the Hudson river. Of course, a flock of sufficiently large drones could cause similar damage. However, flocks of drones are rare; they are typically operated individually. In addition, most consumer drones, such as the DJI Phantom, are only one-third the weight of an average Canada goose. I am not aware of any engine ingestion testing of the Phantom, but it is certainly conceivable that it could cause some degree of damage to an engine but likely not of the severity of what occurred with the US Airways flight.

What is the likelihood of a mid-air collision involving a drone? In order for a collision to occur, the drone needs to be at the same altitude and in the same location as another aircraft. An analysis of radar data indicates that there are large portions of the United States where the risk of encountering another aircraft is negligible at the altitudes a small drone is capable of flying. However, there are portions of our airspace where the likelihood of a collision is orders of magnitude more significant.

Since it is in their interest to prevent accidents, some drone manufacturers, but not all, have been implementing altitude limits and geofencing. Geofencing uses GPS location to help prevent drones from flying in certain geographic areas, such as near airports and over national parks. For some drones, altitude limits and geofencing may be hard constraints representing areas where the drone will not function. For other drones, these constraints are presented as warnings and can be overridden by the operator. In addition to integrating altitude limitations and geofencing into drones, the drone industry has been working with the FAA to educate the public on the safe operation of drones.

In assessing the likelihood of collision with these mitigations in place, it is helpful to break the risk down into four categories of user: the conscientious user, the naïve user, the reckless user, and the bad actor. I will not discuss the bad actor here because this category is unlikely to be significantly impacted by legislation, though I do appreciate that there is a threat it could provide in the future.

First, the conscientious user is one who strives to operate within the altitude limits and geofences and diligently follows FAA guidance and regulation. For this category of user, the likelihood of collision can be quite low. The altitude is measured by a barometer. Failures in the barometer are exceptionally rare. Errors in the estimation of the geographic location are much more common. However, the system can be engineered in a way to help prevent unexpected "fly aways" where the aircraft departs the operating region. For example, in the event of loss of or errors in GPS or the radio control system, the system can be programed to either return to base or to land slowly. In the drones I am aware of, they are designed to only decrease altitude, never increase altitude, in the case of a fault. The likelihood of different failure modes might be estimated to some extent from service call data from vendors.

Second, the naïve user is one who may not be aware of the FAA guidance and regulations. They may receive their drone as a gift and may have noticed something about "know before you fly" packaged up with their drone, but they have no clue on how to estimate an altitude of 400 feet and they might not be aware that they are within a couple miles of a small general aviation airport. If their drone does not have at least a default altitude limit or geofencing, this can result in a safety risk far beyond that posed by the conscientious user. For decades, aircraft modelers and enthusiasts have maintained an exceptional safety record and the community has largely consisted of conscientious users. However, with the mass production of prebuilt consumer drones, many believe that the composition of the community is likely to change and will include a greater proportion of naïve users. The risk presented by these users may be mitigated by education and technology features that provide pertinent information at the time of operation.

Third, the reckless user is one who is aware of regulations but, for example, cares more about capturing a photograph or video than the safety of others. They are not bad actors who are acting primarily to harm others, but they can still pose significant risk. They might ignore or disable altitude-limits or geofences if made easy by the manufacturer, but they generally would not have the capability to reprogram the firmware or modify the hardware to override safety features.

It may be tempting to use the FAA database of alleged pilot sightings of drones to arrive at a quantitative measure of risk. However, as the Academy of Model Aeronautics has observed, many of the sightings are likely military aircraft or recreational aircraft operated legally and safely. It is also possible that some of the sightings were birds. I suspect that some of the pilot sightings involve drones flown by reckless users, but it is difficult to tell since the database does not include details of follow-up investigation. In addition to the potential for the database to lead to an overestimate of risk, there is also the potential to underestimate risk. Many drones are very difficult to see. As a pilot, I know that even conventional manned aircraft can be difficult to see, and so it is possible that there have been potential incidents that went unreported. Although the database of reports is helpful in drawing attention to potential hazards, it does not allow us to precisely measure risk.

Now, the second question: what are the best technologies and policies to mitigate these risks?

There are several technologies to help mitigate risk. I have already mentioned altitude limitations and geofencing. Altitude limits can be implemented fairly reliably and only require a barometric altimeter. A barometric altimeter measures pressure altitude at takeoff and the drone's software can make sure that the altitude of the vehicle not exceed some threshold above that. Of course, if you take off at the top of a mountain, you may be able to go to an area where you are well over 400 feet above the ground. There is not much you can do about this unless your drone is equipped with a terrain database or a laser altimeter. Implementing geofencing is even more difficult because it requires an up-to-date database of geofenced locations and accurate GPS position information, but the safety risk can be significantly reduced with such technology.

Altitude limitation and geofencing technologies are available today and implemented on many of the commercial drones, but should they be mandated? If so, in what form? I am not a policymaker, but I believe it would be wise to establish default altitude limits for non-toy drones capable of flying above 400 feet. The cost to add this safety feature if it does not already exist is fairly negligible. Most drones capable of flying above 400 feet already have an altimeter. One of the first things a new user might be tempted to do after opening the box is see how high the drone can go. A default altitude limit will not ensure safety or prevent interference with firefighting on its own, but it will certainly help naïve users and discourage reckless users. However, prohibiting a conscientious user from overriding the altitude limit is problematic. For example, authorized users such as firefighters might need to go above 400 feet in order to collect imagery of the wildfire they are fighting. The exact approach for overriding limits is still being thought through by industry, and I believe it is too early to mandate a particular mechanism. You do not want it to be especially easy for a casual user to dismiss altitude limits. An altitude limit for an appropriate category of aircraft that can be overridden in some way by a competent operator can go a long way in reducing risk.

Geofencing is more complicated and there are many cases where you may want it disabled. For example, some of my students were authorized to fly drones within a netted cage right next to Moffett Federal Airfield during a convention at NASA Ames on unmanned aircraft. If a geofence were strictly mandated, they would not be able to perform their demonstration. Similarly, fire fighters within an emergency temporary flight restriction should have access to drone technology. Appropriate users need the ability to override geofences. How to determine whether a user should be permitted to override a geofence is unclear and raises complicated questions. A basic geofencing capability implemented by drone manufacturers with the capability of user overrides would help prevent unintentional airspace violations.

Altitude limitations and geofences are near-term risk mitigation measures, but it is becoming clear that some kind of infrastructure will be needed to facilitate the integration of commercial drones into the airspace to support applications such as goods delivery and agricultural monitoring. NASA Ames, in collaboration with industry and academia, has been pursuing the development of the UAS Traffic Management System. I believe this is a strong path forward in bringing together many of the features of air traffic control to ensure safe and efficient drone operation. However, there is still tremendous research to be done in terms of security, resource allocation, and contingency management.

When flying in the same airspace as manned aircraft, a "sense and avoid" system is likely to be necessary to help prevent collision. The FAA has successfully flight tested a version of ACAS X on a Predator B aircraft. The Predator B is equipped with much more sophisticated surveillance systems and avionics than can be placed on smaller, far less expensive drones. It is likely, however, that much of the technology can be adapted for sensor systems that are appropriate for smaller drones. I anticipate that sense and avoid technology will progress significantly over the next few years. Estimating the reduction of risk with such sense and avoid systems can be done through modeling and simulation.

In conclusion, I offer the following: The growing popularity of commercially available drones presents a risk that should not be ignored. Education should play a major role in risk reduction. In addition, there are technologies that can be easily implemented by drone manufacturers on the relevant categories of drones to help prevent inadvertent altitude or geographic airspace violations. It is in the interest of the drone industry as a whole to implement these safety measures. It is in the interest of our nation to support the research needed to ensure aviation safety as our technology evolves. The future of unmanned aviation is bright and has the potential to flourish into an entirely new industry with applications yet to be imagined.

Thank you for this opportunity, and I am happy to be a resource to this subcommittee.

In: Unmanned Aerial Systems
Editor: Wilfred Sparks

ISBN: 978-1-63484-652-3

Chapter 12

PRESIDENTIAL MEMORANDUM: PROMOTING ECONOMIC COMPETITIVENESS WHILE SAFEGUARDING PRIVACY, CIVIL RIGHTS, AND CIVIL LIBERTIES IN DOMESTIC USE OF UNMANNED AIRCRAFT SYSTEMS*

Barack Obama

MEMORANDUM FOR THE HEADS OF EXECUTIVE DEPARTMENTS AND AGENCIES

SUBJECT: Promoting Economic Competitiveness While Safeguarding Privacy, Civil Rights, and Civil Liberties in Domestic Use of Unmanned Aircraft Systems

Unmanned Aircraft Systems (UAS) technology continues to improve rapidly, and increasingly UAS are able to perform a variety of missions with greater operational flexibility and at a lower cost than comparable manned aircraft. A wide spectrum of domestic users -- including industry, private citizens, and Federal, State, local, tribal, and territorial governments -- are

---

* This is an edited, reformatted and augmented version of a presidential memorandum issued by the White House, Office of the Press Secretary, February 15, 2015.

using or expect to use these systems, which may play a transformative role in fields as diverse as urban infrastructure management, farming, public safety, coastal security, military training, search and rescue, and disaster response.

The Congress recognized the potential wide-ranging benefits of UAS operations within the United States in the FAA Modernization and Reform Act of 2012 (Public Law 112-95), which requires a plan to safely integrate civil UAS into the National Airspace System (NAS) by September 30, 2015. As compared to manned aircraft, UAS may provide lower-cost operation and augment existing capabilities while reducing risks to human life. Estimates suggest the positive economic impact to U.S. industry of the integration of UAS into the NAS could be substantial and likely will grow for the foreseeable future.

As UAS are integrated into the NAS, the Federal Government will take steps to ensure that the integration takes into account not only our economic competitiveness and public safety, but also the privacy, civil rights, and civil liberties concerns these systems may raise.

By the authority vested in me as President by the Constitution and the laws of the United States of America, and in order to establish transparent principles that govern the Federal Government's use of UAS in the NAS, and to promote the responsible use of this technology in the private and commercial sectors, it is hereby ordered as follows:

Section 1. UAS Policies and Procedures for Federal Government Use. The Federal Government currently operates UAS in the United States for several purposes, including to manage Federal lands, monitor wildfires, conduct scientific research, monitor our borders, support law enforcement, and effectively train our military. As with information collected by the Federal Government using any technology, where UAS is the platform for collection, information must be collected, used, retained, and disseminated consistent with the Constitution, Federal law, and other applicable regulations and policies. Agencies must, for example, comply with the Privacy Act of 1974 (5 U.S.C. 552a) (the "Privacy Act"), which, among other things, restricts the collection and dissemination of individuals' information that is maintained in systems of records, including personally identifiable information (PII), and permits individuals to seek access to and amendment of records.

(a) Privacy Protections. Particularly in light of the diverse potential uses of UAS in the NAS, expected advancements in UAS technologies, and the anticipated increase in UAS use in the future, the Federal Government shall take steps to ensure that privacy protections and policies relative to UAS continue to keep pace with these developments. Accordingly, agencies shall,

prior to deployment of new UAS technology and at least every 3 years, examine their existing UAS policies and procedures relating to the collection, use, retention, and dissemination of information obtained by UAS, to ensure that privacy, civil rights, and civil liberties are protected. Agencies shall update their policies and procedures, or issue new policies and procedures, as necessary. In addition to requiring compliance with the Privacy Act in applicable circumstances, agencies that collect information through UAS in the NAS shall ensure that their policies and procedures with respect to such information incorporate the following requirements:

(i) Collection and Use. Agencies shall only collect information using UAS, or use UAS-collected information, to the extent that such collection or use is consistent with and relevant to an authorized purpose.

(ii) Retention. Information collected using UAS that may contain PII shall not be retained for more than 180 days unless retention of the information is determined to be necessary to an authorized mission of the retaining agency, is maintained in a system of records covered by the Privacy Act, or is required to be retained for a longer period by any other applicable law or regulation.

(iii) Dissemination. UAS-collected information that is not maintained in a system of records covered by the Privacy Act shall not be disseminated outside of the agency unless dissemination is required by law, or fulfills an authorized purpose and complies with agency requirements.

(b) Civil Rights and Civil Liberties Protections. To protect civil rights and civil liberties, agencies shall:

(i) ensure that policies are in place to prohibit the collection, use, retention, or dissemination of data in any manner that would violate the First Amendment or in any manner that would discriminate against persons based upon their ethnicity, race, gender, national origin, religion, sexual orientation, or gender identity, in violation of law;

(ii) ensure that UAS activities are performed in a manner consistent with the Constitution and applicable laws, Executive Orders, and other Presidential directives; and

(iii) ensure that adequate procedures are in place to receive, investigate, and address, as appropriate, privacy, civil rights, and civil liberties complaints.

(c) Accountability. To provide for effective oversight, agencies shall:

(i) ensure that oversight procedures for agencies' UAS use, including audits or assessments, comply with existing agency policies and regulations;

(ii) verify the existence of rules of conduct and training for Federal Government personnel and contractors who work on UAS programs, and procedures for reporting suspected cases of misuse or abuse of UAS technologies;

(iii) establish policies and procedures, or confirm that policies and procedures are in place, that provide meaningful oversight of individuals who have access to sensitive information (including any PII) collected using UAS;

(iv) ensure that any data-sharing agreements or policies, data use policies, and record management policies applicable to UAS conform to applicable laws, regulations, and policies;

(v) establish policies and procedures, or confirm that policies and procedures are in place, to authorize the use of UAS in response to a request for UAS assistance in support of Federal, State, local, tribal, or territorial government operations; and

(vi) require that State, local, tribal, and territorial government recipients of Federal grant funding for the purchase or use of UAS for their own operations have in place policies and procedures to safeguard individuals' privacy, civil rights, and civil liberties prior to expending such funds.

(d) Transparency. To promote transparency about their UAS activities within the NAS, agencies that use UAS shall, while not revealing information that could reasonably be expected to compromise law enforcement or national security:

(i) provide notice to the public regarding where the agency's UAS are authorized to operate in the NAS;

(ii) keep the public informed about the agency's UAS program as well as changes that would significantly affect privacy, civil rights, or civil liberties; and

(iii) make available to the public, on an annual basis, a general summary of the agency's UAS operations during the previous fiscal year, to include a brief description of types or categories of missions flown, and the number of times the agency provided assistance to other agencies, or to State, local, tribal, or territorial governments.

(e) Reports. Within 180 days of the date of this memorandum, agencies shall provide the President with a status report on the implementation of this section. Within 1 year of the date of this memorandum, agencies shall publish information on how to access their publicly available policies and procedures implementing this section.

Sec. 2. Multi-stakeholder Engagement Process. In addition to the Federal uses of UAS described in section 1 of this memorandum, the combination of greater operational flexibility, lower capital requirements, and lower operating costs could allow UAS to be a transformative technology in the commercial and private sectors for fields as diverse as urban infrastructure management, farming, and disaster response. Although these opportunities will enhance American economic competitiveness, our Nation must be mindful of the potential implications for privacy, civil rights, and civil liberties. The Federal Government is committed to promoting the responsible use of this technology in a way that does not diminish rights and freedoms.

(a) There is hereby established a multi-stakeholder engagement process to develop and communicate best practices for privacy, accountability, and transparency issues regarding commercial and private UAS use in the NAS. The process will include stakeholders from the private sector.

(b) Within 90 days of the date of this memorandum, the Department of Commerce, through the National Telecommunications and Information Administration, and in consultation with other interested agencies, will initiate this multi-stakeholder engagement process to develop a framework regarding privacy, accountability, and transparency for commercial and private UAS use. For this process, commercial and private use includes the use of UAS for commercial purposes as civil aircraft, even if the use would qualify a UAS as a public aircraft under 49 U.S.C. 40102(a)(41) and 40125. The process shall not focus on law enforcement or other noncommercial governmental use.

Sec. 3. Definitions. As used in this memorandum:

(a) "Agencies" means executive departments and agencies of the Federal Government that conduct UAS operations in the NAS.

(b) "Federal Government use" means operations in which agencies operate UAS in the NAS. Federal Government use includes agency UAS operations on behalf of another agency or on behalf of a State, local, tribal, or territorial government, or when a nongovernmental entity operates UAS on behalf of an agency.

(c) "National Airspace System" means the common network of U.S. airspace; air navigation facilities, equipment, and services; airports or landing areas; aeronautical charts, information, and services; related rules, regulations, and procedures; technical information; and manpower and material. Included in this definition are system components shared jointly by the Departments of Defense, Transportation, and Homeland Security.

(d) "Unmanned Aircraft System" means an unmanned aircraft (an aircraft that is operated without direct human intervention from within or on the aircraft) and associated elements (including communication links and components that control the unmanned aircraft) that are required for the pilot or system operator in command to operate safely and efficiently in the NAS.

(e) "Personally identifiable information" refers to information that can be used to distinguish or trace an individual's identity, either alone or when combined with other personal or identifying information that is linked or linkable to a specific individual, as set forth in Office of Management and Budget Memorandum M-07-16 (May 22, 2007) and Office of Management and Budget Memorandum M-10-23 (June 25, 2010).

Sec. 4. General Provisions.

(a) This memorandum complements and is not intended to supersede existing laws and policies for UAS operations in the NAS, including the National Strategy for Aviation Security and its supporting plans, the FAA Modernization and Reform Act of 2012, the Federal Aviation Administration's (FAA's) Integration of Civil UAS in the NAS Roadmap, and the FAA's UAS Comprehensive Plan.

(b) This memorandum shall be implemented consistent with applicable law, and subject to the availability of appropriations.

(c) Nothing in this memorandum shall be construed to impair or otherwise affect:

(i) the authority granted by law to an executive department, agency, or the head thereof; or

(ii) the functions of the Director of the Office of Management and Budget relating to budgetary, administrative, or legislative proposals.

(d) Independent agencies are strongly encouraged to comply with this memorandum.

(e) This memorandum is not intended to, and does not, create any right or benefit, substantive or procedural, enforceable at law or in equity by any party against the United States, its departments, agencies, or entities, its officers, employees, or agents, or any other person.

(f) The Secretary of Commerce is hereby authorized and directed to publish this memorandum in the Federal Register.

BARACK OBAMA

INDEX

#

21st century, 64

A

abuse, 166
access, 8, 10, 11, 15, 18, 19, 21, 23, 24, 27, 28, 33, 42, 47, 52, 62, 85, 117, 133, 148, 161, 164, 166
accommodation, 12, 18
accountability, 60, 61, 80, 128, 167
adaptation, 37
advancements, 88, 164
adverse weather, 95
aerospace, 55, 91, 98
age, 64, 123, 124
agencies, 7, 21, 28, 37, 39, 41, 42, 47, 51, 53, 59, 60, 61, 70, 77, 85, 89, 90, 93, 96, 97, 103, 106, 118, 142, 164, 165, 166, 167, 168
agricultural producers, 83
agriculture, 19, 26, 30, 84, 85, 88, 94
Air Force, 44, 140, 146, 157
airports, 19, 24, 89, 95, 97, 114, 115, 116, 120, 124, 135, 159, 167
air-traffic, 7
airways, 64
Alaska, 21, 39, 44, 46, 51, 73, 146
appointments, 57
appropriations, 168
ARC, 52, 55, 98
asset(s), 51, 65, 97, 101, 102, 104, 105
atmosphere, 17, 62
audit(s), 6, 40, 165
authorit(ies), 2, 5, 16, 17, 23, 24, 27, 28, 30, 34, 40, 44, 45, 46, 50, 52, 77, 80, 90, 123, 97, 118, 123, 128, 129, 132, 136, 142, 153, 164, 168
automation, 45, 76, 117
automobiles, 76
avoidance, 117, 119, 120, 122, 135, 158
awareness, 5, 37, 55, 95, 96, 98, 104, 128

B

background information, 38
bad behavior, 136
ban, 12, 37
barriers, 51, 118
base, 105, 113, 140, 159
batteries, 115
benefits, 25, 55, 69, 76, 77, 98, 118, 119, 128, 164
bias, 147
birds, 90, 131, 136, 146, 147, 160
Bureau of Land Management, 133, 149
business model, 71
businesses, 84
buyers, 76

C

campaigns, 106
career development, 130
case studies, 40
category a, 149
cell phones, 65
certificate, 16, 18, 23, 25, 27, 31, 33, 35, 46, 50, 51, 52, 123, 154
certification, 3, 15, 16, 24, 31, 32, 33, 34, 43, 46, 53, 55, 59, 98, 117, 118, 122, 125
CFR, 46
challenges, 2, 5, 25, 28, 29, 36, 37, 38, 44, 47, 50, 51, 53, 55, 76, 83, 94, 98, 101, 116, 125, 128
Chicago, 144
Chief Justice, 65
cities, 20
citizens, 163
civil liberties, 60, 72, 164, 165, 166, 167
civil rights, 72, 164, 165, 166, 167
clarity, 115
classes, 61
classification, 17, 72
climate, 18, 71
CNN, 28, 55, 98
Coast Guard, 142, 149
collaboration, 22, 46, 95, 118, 162
collisions, 29, 44, 77, 91, 105, 119, 131, 136, 138, 152
color, 96
commercial, vii, 1, 2, 3, 4, 5, 6, 10, 11, 12, 19, 30, 31, 33, 35, 37, 38, 40, 41, 50, 51, 59, 60, 61, 64, 68, 71, 76, 77, 78, 79, 80, 81, 84, 88, 89, 90, 93, 94, 97, 99, 110, 111, 112, 113, 114, 116, 117, 118, 119, 121, 122, 123, 124, 128, 130, 131, 132, 136, 140, 142, 143, 147, 151, 152, 154, 158, 161, 162, 164, 166, 167
common law, 63, 67
common sense, 95
communication, 15, 73, 103, 104, 118, 119, 132, 150, 153, 167
communit(ies), 41, 45, 52, 54, 58, 71, 85, 89, 104, 106, 121, 127, 128, 129, 132, 133, 135, 140, 143, 154, 160
compensation, 26, 123, 154
competition, 127
competitive advantage, 124
competitiveness, 72
compilation, 89
complexity, 31, 37, 103
compliance, 38, 45, 46, 55, 118, 129, 145, 165
composition, 160
computer, 70
computing, 88
conference, 22, 77
Congress, 5, 8, 43, 50, 53, 56, 58, 59, 71, 80, 88, 93, 122, 123, 124, 129, 143, 164
consensus, 8, 22, 46, 95
conservation, 128
Constitution, 164, 165
construction, 22, 45, 142
consumers, 90, 136, 142
contingency, 134, 162
convention, 161
conversations, 131, 151
cooperation, 26, 95
cooperative agreements, 10
coordination, 51, 97, 101, 102, 107, 132, 133, 150, 153
cost, 15, 29, 70, 85, 161, 163, 164
covering, 27
criminal activity, 65
criminal statutes, 67
critical infrastructure, 135, 136, 148
crop(s), 4, 20, 83, 84, 85, 98
cues, 44, 134
culture, 93, 95
customers, 28, 75, 76, 129
Customs and Border Protection, 7, 44

D

danger, 145, 150
data analysis, 54
data collection, 54, 85

data gathering, 6
data set, 143, 144, 145
data transfer, 54
database, 113, 148, 160, 161
decision-making process, 55
deficiency, 26
demonstrations, 7, 11
Denmark, 26
Department of Agriculture, 107
Department of Commerce, 39, 44, 167
Department of Defense, 3, 5, 7, 8, 39
Department of Homeland Security (DHS), 3, 5, 7, 29, 39, 44, 107, 157
Department of Justice, 41, 46
Department of Transportation (DOT), 2, 3, 13, 37, 45, 53
designers, 14
detection, 90, 134
directives, 165
disaster, 4, 26, 71, 94, 164, 167
disaster assistance, 4
disclosure, 67
distribution, 76
divergence, 65
diversity, 52, 103, 158
DOI, 101, 102, 103, 106
draft, 37, 41, 78, 154
drawing, 160
drought, 96, 102
drugs, 95

E

economic competitiveness, 164, 167
education, 37, 90, 94, 95, 96, 106, 127, 132, 133, 136, 149, 154, 160
educational materials, 149
educational programs, 129
emergency, 26, 52, 94, 107, 158, 161
emergency response, 94
employees, 17, 46, 112, 168
employment, 78
endurance, 28
energy, 71, 88
enforcement, 16, 17, 45, 46, 55, 65, 89, 90, 96, 124, 128, 132, 137, 150, 153
engineering, 104, 130
environment(s), 37, 50, 70, 84, 85, 95, 97, 107, 129, 130
environmental conditions, 83
environmental impact, 38, 84, 85
environmental organizations, 85
equipment, 14, 51, 102, 104, 122, 136, 167
equity, 168
ethnicity, 165
Europe, 30, 77, 78, 79
European Commission, 77, 78
European Union, 46
evidence, 6, 40, 65, 96, 147
evolution, 12
Executive Order, 165

F

FAA, vii, 1, 2, 3, 4, 5, 6, 7, 8, 10, 11, 12, 13, 14, 15, 16, 17, 18, 19, 20, 21, 22, 23, 24, 25, 27, 28, 29, 30, 31, 32, 34, 35, 37, 38, 39, 40, 41, 42, 43, 44, 45, 46, 47, 49, 50, 51, 52, 53, 54, 55, 56, 58, 59, 71, 73, 77, 78, 79, 80, 81, 85, 87, 88, 89, 90, 91, 93, 94, 95, 96, 97, 98, 99, 106, 110, 111, 112, 113, 114, 116, 117, 118, 119, 120, 123, 125, 129, 130, 131, 132, 133, 134, 135, 136, 137, 138, 139, 140, 141, 142, 143, 144, 145, 146, 147, 148, 149, 150, 151, 152, 153, 154, 157, 159, 160, 162, 164, 168
Facebook, 106
Fairbanks, 46
farmers, 37, 83, 84, 85
fear, 85
federal advisory, 56, 99
federal agency, 44, 46
federal government, 29, 44, 59, 60, 61, 93, 164, 165, 167
federal law, 10
Federal Register, 43, 47, 168
fencing, 90, 120, 129
filters, 152

financial, 27
fire detection, 103
fire suppression, 104, 105
fires, 32, 102, 103, 105
First Amendment, 58, 66, 67, 68, 165
first responders, 54
fiscal year, 46, 60, 166
flaws, 91
flexibility, 53, 55, 71, 98, 163, 166
flight(s), 2, 7, 12, 17, 18, 19, 23, 24, 26, 27, 28, 31, 36, 44, 45, 51, 52, 53, 58, 60, 62, 64, 68, 69, 70, 73, 76, 78, 80, 88, 89, 90, 97, 103, 114, 115, 116, 117, 119, 121, 122, 128, 129, 132, 134, 145, 147, 148, 149, 152, 158, 161, 162
food, 85
food production, 85
football, 4, 54, 90, 96, 148
force, 61
forest fire, 71, 89
Fourth Amendment, 63, 64, 65, 66
framing, 62
France, 2, 3, 5, 31, 32, 33, 35, 40
freedom, 58, 67
funding, 2, 10, 18, 25, 27, 29, 72, 120, 122, 124, 166
funds, 15, 72, 166

G

GAO, 1, 2, 3, 5, 8, 9, 11, 14, 16, 18, 19, 20, 24, 26, 34, 43, 44, 46
gender identity, 165
geography, 18
Georgia, 28, 29, 40, 146
goose, 159
governments, 8, 60, 95, 163, 166
GPS, 65, 129, 134, 136, 159, 161
grants, 10, 28, 29, 97
grasslands, 103
group work, 25
growth, 35, 89, 105, 110, 112, 114, 116, 124, 128
guidance, 2, 13, 15, 21, 22, 26, 40, 42, 45, 54, 103, 116, 118, 159, 160
guidelines, 35, 37, 45, 89, 123, 129, 131, 133, 136, 143, 144, 145, 151
guilty, 145

H

harmonization, 30, 78, 80, 120
Hawaii, 44
hazardous materials, 91
hazards, 111, 158, 160
health, 103
height, 31, 33, 62
history, 134
hobby, 17, 45, 88, 96, 110, 113, 114, 116, 123, 127, 129, 130, 133, 149
homes, 65
host, 142
hot spots, 31
House, 72, 87, 93, 101, 109, 127, 157
House Committee on Transportation and Infrastructure, 87, 93, 101, 109, 127, 157
human, 10, 44, 46, 69, 70, 73, 77, 94, 104, 115, 119, 120, 164, 167
humidity, 78

I

Iceland, 27, 44
identification, 143
identity, 90, 123, 168
illumination, 147
image(s), 62, 68, 154
imagery, 59, 84, 161
imagination, 94
imprisonment, 91
improvements, 35, 62
India, 37
individuals, 20, 36, 46, 59, 66, 71, 72, 90, 107, 130, 164, 166
industr(ies), 2, 3, 4, 5, 6, 8, 10, 13, 18, 19, 20, 21, 22, 23, 24, 25, 26, 27, 29, 30, 35, 38, 39, 46, 50, 51, 53, 54, 55, 58, 78, 79, 84, 88, 90, 94, 95, 96, 97, 98, 106, 110,

112, 113, 115, 122, 124, 128, 133, 159, 161, 162, 163, 164
infestations, 84
infrastructure, 15, 94, 95, 98, 129, 162, 164, 167
ingestion, 158
injur(ies), 90, 105, 128
inspections, 6, 88, 104
institutions, 10, 11, 25, 46, 54, 118
integration, vii, 1, 2, 3, 4, 5, 7, 8, 10, 12, 13, 15, 18, 20, 21, 22, 24, 25, 26, 28, 29, 30, 32, 36, 37, 38, 39, 42, 43, 44, 45, 47, 49, 50, 51, 52, 53, 54, 55, 58, 59, 68, 69, 76, 77, 88, 93, 94, 95, 96, 97, 98, 103, 104, 106, 110, 112, 116, 118, 119, 120, 121, 122, 162, 164
integrity, 44, 54
intelligence, 31, 77
interface, 96
interference, 91, 132, 158, 161
international standards, 44, 47
interoperability, 51, 97, 121
intervention, 167
invasion of privacy, 63, 66, 67, 68, 73
investment, 25
issues, 3, 8, 16, 22, 25, 29, 36, 37, 38, 39, 40, 42, 46, 54, 61, 87, 94, 97, 107, 109, 119, 128, 138, 139, 148, 153, 167

J

Japan, 30, 46
jurisdiction, 102

K

knots, 17

L

landscape, 53
latency, 119
Latvia, 77
law enforcement, 10, 11, 37, 52, 54, 59, 64, 65, 88, 89, 96, 123, 132, 133, 137, 142, 150, 153, 164, 166, 167
law enforcement organizations, 10, 11
laws, 58, 59, 60, 62, 63, 67, 90, 96, 132, 153, 164, 165, 166, 168
leadership, 58, 77, 96, 127, 129
learning, 150
legislation, 58, 61, 72, 88, 159
legislative proposals, 168
liability insurance, 34
light, 60, 63, 65, 140, 146, 147, 152, 153, 164
livestock, 83
local authorities, 132
local government, 8, 10, 11, 25, 46, 60, 72
love, 76
low risk, 32, 53, 76, 84, 97

M

magnitude, 131, 159
Major League Baseball, 148
majority, 6, 60, 64, 65, 66, 73, 90, 94, 114, 130, 150
mammal, 51
management, 29, 46, 68, 69, 84, 85, 104, 162, 164, 166, 167
manpower, 167
manufacturing, 7
mapping, 104
marijuana, 64, 65
marketing, 142
Maryland, 44
mass, 160
materials, 15, 17, 133, 134, 136
mathematics, 130
matter, 144
media, 72, 88, 91, 95, 106, 130, 131, 139, 149, 150, 151, 153, 158
medical, 115
membership, 129
memorandums of understanding, 22
methodology, 6, 38
Mexico, 28, 40, 46

Miami, 144, 148
military, 2, 4, 6, 10, 40, 45, 88, 90, 102, 118, 130, 131, 135, 136, 140, 141, 143, 151, 152, 158, 160, 164
mission(s), 6, 7, 17, 60, 90, 103, 104, 118, 163, 165, 166
misuse, 95, 166
mobile phone, 65
models, 157
modernization, 46
modifications, 61
moisture, 84
Montana, 46

N

naming, 10
narcotics, 65
NAS, 45, 49, 50, 51, 52, 53, 54, 55, 56, 60, 68, 93, 94, 97, 98, 99, 109, 110, 111, 113, 114, 116, 117, 118, 119, 120, 121, 122, 123, 124, 164, 166, 167, 168
National Aeronautics and Space Administration, 5, 7, 39, 51, 97
national airspace, vii, 1, 2, 4, 5, 6, 7, 8, 10, 11, 12, 13, 14, 15, 16, 17, 18, 21, 24, 26, 30, 32, 36, 38, 39, 40, 41, 42, 43, 45, 46, 59, 109, 123, 132, 150, 153
National Defense Authorization Act, 8
National Environmental Policy Act, 38
National Football League, 148
National Forest System, 103
National Oceanic and Atmospheric Administration, 41, 46
national origin, 165
national parks, 159
national security, 50, 166
National Strategy, 168
natural disaster, 124
natural resource management, 104
natural resources, 102, 104
Natural Resources Conservation Service (NRCS), 86
neutral, 67
news coverage, 148
next generation, 158
Next Generation Air Transportation System, 7, 55, 98
NextGen, 7, 55, 73, 98, 120, 124
NFL, 54, 96
North America, 77
nutrient(s), 84

O

Obama, President Barack, 60, 163
obstacles, 3, 7, 119
Office of Management and Budget, 13, 168
officials, 2, 5, 17, 18, 22, 25, 26, 29, 31, 35, 39, 77
oil, 88
omission, 138
operating costs, 166
operations, vii, 1, 2, 3, 4, 5, 6, 10, 11, 12, 13, 14, 15, 16, 17, 18, 19, 20, 21, 23, 24, 25, 27, 29, 30, 31, 32, 33, 34, 35, 36, 38, 39, 45, 46, 50, 51, 52, 53, 54, 55, 58, 60, 71, 72, 73, 76, 77, 78, 79, 80, 84, 88, 89, 90, 94, 96, 97, 98, 102, 105, 106, 107, 110, 111, 112, 114, 117, 118, 119, 120, 122, 123, 124, 125, 128, 132, 137, 140, 142, 143, 149, 150, 153, 158, 164, 166, 167, 168
opportunities, 22, 53, 55, 75, 76, 77, 85, 88, 98, 107, 167
optimism, 64
outreach, 54, 94, 95, 96, 106
oversight, 26, 46, 132, 154, 165, 166
ownership, 67, 85

P

Pacific, 27, 102
Parliament, 35, 36
participants, 46
penalties, 89, 124, 132, 136, 153
permission, 32, 33, 52, 76, 78, 79, 85
permit, 52, 79, 88
pesticide, 30, 85

pests, 84
photographs, 6, 64
pipeline, 6, 58, 71, 88, 112
plants, 64
platform, 164
playing, 149
policy, 7, 13, 18, 19, 21, 43, 91, 103, 128, 131, 151, 153
policy issues, 21
population, 110
portfolio, 18
power plants, 95, 136, 149
precedents, 63
precipitation, 78
preservation, 128
President, 60, 109, 149, 164, 166
primary function, 120
principles, 164
private information, 66
private sector, 7, 26, 29, 54, 60, 90, 167
professionals, 130
profit, 8, 46, 56, 99
profitability, 84
programming, 45, 89, 132, 143, 154
project, 20, 42, 107
proliferation, 89, 130, 158
property rights, 62
proposed regulations, 32
protection, 46, 58, 63, 65, 84, 127
prototype, 51, 98
public awareness, 54
public concern(s), 21
public education, 106
public opinion, 139
public safety, 41, 94, 103, 123, 139, 164
public sector, 54
public service, 96, 106, 133, 142
publishing, 22, 110
Puerto Rico, 83, 102
PUMA, 51

Q

qualifications, 34, 111
quality of life, 128

R

race, 124, 165
radar, 68, 124, 159
radio, 8, 10, 44, 73, 103, 104, 119, 134, 140, 159
radius, 149
real estate, 4, 6, 19
real time, 104
reality, 117
recall, 158
reception, 119
recognition, 94
recommendations, 8, 17, 22, 31, 43, 52, 55, 98, 104, 111, 132, 137, 151
recreation, 94, 120, 123
recreational, 17, 21, 35, 36, 45, 88, 89, 94, 101, 103, 112, 123, 127, 128, 129, 131, 133, 136, 140, 142, 143, 144, 145, 151, 154, 160
redundancy, 134
Reform, vii, 1, 4, 5, 38, 39, 41, 42, 43, 50, 53, 59, 88, 94, 164, 168
regulations, 2, 3, 4, 5, 10, 12, 13, 14, 17, 26, 30, 31, 32, 34, 35, 37, 38, 39, 40, 46, 50, 70, 72, 85, 88, 95, 111, 114, 117, 120, 122, 123, 124, 128, 132, 133, 160, 164, 165, 166, 167
regulatory agencies, 85
regulatory framework, 49, 50, 71, 77, 78, 79, 80, 94, 110, 128, 143
regulatory requirements, 2, 6, 34, 40
relevance, 66
reliability, 21, 34, 134
relief, 50
religion, 165
religious beliefs, 66
remote sensing, 78
requirement(s), 10, 12, 15, 16, 17, 18, 24, 27, 31, 33, 34, 41, 42, 43, 46, 59, 77, 117, 118, 119, 120, 121, 122, 123, 143, 165, 166
researchers, 36
residential neighborhood, 140
resolution, 68

resource allocation, 162
resource management, 104
resources, vii, 1, 5, 13, 15, 18, 25, 28, 29, 36, 38, 39, 40, 50, 55, 98, 101, 102, 103, 104, 114, 124
response, vii, 1, 10, 12, 26, 36, 65, 71, 89, 101, 102, 103, 105, 137, 138, 164, 166, 167
response capacity, 102
restrictions, 30, 31, 32, 33, 66, 111, 124, 129, 139, 148, 149, 152
retail, 133, 135
revenue, 18, 23, 24, 27
rights, 32, 85, 167
risk(s), 15, 30, 31, 36, 37, 50, 53, 55, 67, 69, 77, 78, 90, 91, 95, 98, 102, 105, 110, 111, 112, 113, 118, 129, 131, 132, 134, 137, 138, 140, 151, 152, 153, 157, 158, 159, 160, 161, 162, 164
risk assessment, 30
robotics, 58
routes, vii, 1, 44, 51, 104
rules, 3, 13, 17, 20, 30, 31, 32, 36, 45, 53, 59, 61, 62, 71, 76, 77, 78, 79, 80, 85, 95, 96, 111, 112, 119, 125, 132, 133, 136, 143, 153, 154, 165, 167
rural areas, 98

S

safety, vii, 1, 2, 4, 7, 14, 15, 16, 17, 20, 21, 23, 25, 29, 30, 31, 36, 42, 44, 45, 46, 49, 50, 51, 52, 53, 55, 69, 76, 77, 87, 89, 90, 91, 94, 95, 96, 97, 98, 101, 103, 104, 106, 107, 109, 110, 111, 114, 115, 116, 117, 118, 119, 120, 121, 122, 123, 124, 125, 127, 128, 129, 130, 131, 132, 133, 135, 136, 137, 138, 139, 140, 143, 149, 151, 152, 153, 154, 157, 158, 160, 161, 162
safety zones, 104
scarce resources, 31
science, 124, 130
scope, 2, 6, 38, 45, 68
sea level, 141, 144, 145, 152
Secretary of Commerce, 168
Secretary of Defense, 8
Secretary of Homeland Security, 8
security, 4, 44, 85, 91, 149, 162, 164
seed, 85
semi-structured interviews, 5, 39
Senate, 49, 57, 75, 83
sensor(s), 104, 121, 162
service provider, 7, 14
services, 7, 15, 46, 52, 75, 80, 81, 109, 119, 167
sexual orientation, 165
shape, 115, 147
showing, 70, 134
signals, 44
silver, 141, 146
simulation, 162
society, 64, 76, 128
software, 120, 128, 161
solution, 62, 118, 120, 123, 128
space technology, 99
specifications, 8
spin, 134
stakeholder groups, 38
stakeholders, 2, 5, 7, 8, 13, 15, 21, 30, 40, 51, 90, 95, 97, 102, 118, 122, 131, 133, 139, 151, 152, 153, 167
standardization, 117, 158
state(s), 8, 10, 11, 23, 24, 25, 27, 38, 46, 58, 59, 60, 61, 67, 72, 80, 83, 89, 96, 101, 102, 103, 106, 117, 125, 129, 130, 133, 139, 142, 144
state legislatures, 27, 59, 129
statistics, 60
statutes, 63, 73
statutory authority, 80
structure, 52, 53, 103
style, 113, 141
Super Bowl, 54, 95
supervision, 104
suppression, 101, 102, 105, 107
Supreme Court, 62, 63, 64
surveillance, 6, 26, 72, 80, 162
sustainability, 84

T

target, 117, 122, 124
team members, 46
technical comments, 38
techniques, 84, 129
technological advancement, 124
technolog(ies), 3, 5, 7, 10, 13, 20, 22, 35, 36, 38, 40, 42, 45, 46, 49, 52, 53, 55, 58, 60, 62, 63, 64, 65, 66, 67, 69, 70, 71, 76, 77, 78, 79, 80, 84, 85, 88, 90, 91, 94, 96, 97, 98, 99, 103, 104, 106, 107, 110, 116, 117, 118, 120, 121, 122, 124, 125, 128, 129, 130, 32, 135, 139, 153, 154, 157, 158, 160, 161, 162, 163, 164, 165, 166
temperature, 64, 78, 134
tenants, 128
territorial, 60, 72, 163, 166, 167
testing, 7, 22, 25, 26, 27, 28, 29, 34, 46, 50, 58, 69, 71, 76, 77, 78, 79, 96, 115, 118, 159
thoughts, 102, 158
threats, 84, 129
time frame, 15, 113
time periods, 15
toys, 95
tracks, 68
trade, 5, 38, 90
training, 11, 15, 25, 26, 31, 33, 34, 51, 52, 94, 114, 117, 164, 165
transactions, 44
transference, 120
transformation, 84
transmission, 119
transparency, 60, 61, 80, 145, 153, 166, 167
transport, 46, 103, 115
transportation, 76, 88, 91, 105, 109, 123, 124, 125
transportation infrastructure, 88
Transportation Security Administration, 16
treatment, 84, 85
turbulence, 78

U

U.S. Department of Agriculture (USDA), 84, 86, 107
U.S. Department of Commerce, 46
U.S. Department of the Interior, 46
uniform, 80
United Airlines, 152
United Kingdom, 2, 3, 5, 31, 32, 33, 35, 37, 40
United States (USA), 1, 2, 3, 20, 27, 30, 32, 33, 35, 36, 37, 40, 46, 50, 55, 58, 62, 64, 65, 68, 72, 73, 76, 78, 79, 80, 81, 87, 89, 91, 101, 102, 107, 109, 130, 135, 147, 157, 159, 164, 168
universities, 2, 8, 40, 142
unreasonable searches, 63
urban, 55, 164, 167
urban areas, 55

V

variations, 118
vehicles, 44, 76, 78, 91, 94, 97, 105, 158
venue, 46
Vice President, 75
videos, 106, 114
vision, 6, 36, 96, 98
visual acuity, 115
visual environment, 44
vulnerability, 44, 118

W

waiver, 24, 42
Washington, 44, 45, 47, 95, 102, 113, 119, 135, 139, 141, 146, 148, 149, 152
water, 84, 103, 105
web, 37, 116
Western Australia, 31
White House, 72, 90, 93, 124, 149, 163
wilderness, 70
wildfire, 103, 104, 105, 106, 149, 161
wildland, 101, 102, 104, 107

wildlife, 52
Wisconsin, 149
witnesses, 36, 96
workload, 25, 52
workplace, 130
worldwide, 8, 36, 77

Y

yield, 16, 83, 84, 145